Theodore Gill

Bibliography of the Fishes of the Pacific Coast of the United

states

To the End of 1879

Theodore Gill

Bibliography of the Fishes of the Pacific Coast of the United states
To the End of 1879

ISBN/EAN: 9783337186197

Printed in Europe, USA, Canada, Australia, Japan

Cover: Foto ©berggeist007 / pixelio.de

More available books at **www.hansebooks.com**

BIBLIOGRAPHY

OF THE

FISHES OF THE PACIFIC COAST

OF THE

UNITED STATES

TO

THE END OF 1879.

BY

THEODORE GILL.

———◆———

WASHINGTON:
GOVERNMENT PRINTING OFFICE.
1882.

BIBLIOGRAPHY

OF

THE FISHES OF THE PACIFIC UNITED STATES.

PREFATORY.

The scientific literature relative to the fishes of the western coast of North America is of unusually recent, as well as rapid, growth. Nothing exact was known till the present century had far advanced, for the accounts of the earlier writers, such as Venegas, intead of enlightening the reader, convey absolutely false ideas respecting the character of the ichthyic fauna. Exclusive of incidental notices, the beginnings of an ichthyography of the northwest coast were first published in 1831 (but printed in 1811) in the "Zoographia Rosso-Asiatica" of Pallas; a few species from British Columbia were described by Richardson in 1836, while the fishes of California remained absolutely unknown till 1839, when a glimpse, but an entirely inadequate one, was furnished by Lay and Bennett in their notes and account of species collected during the voyage of the English vessel Blossom. A long silence then supervened, and, with the exceptions thus signalized, and the addition by Storer of a single species of *Syngnathus* in 1846, west-coast ichthyography commenced in 1854 with the announcement, by Professor Agassiz, of the discovery of the remarkable family of Embiotocoids. This was speedily followed by numerous communications, by Dr. Gibbons, Dr. Girard, and Dr. Ayres, on new species of fishes, mostly from the Californian waters, but partly from the Oregonian ones. As early as 1858, nearly 200 species had been made known, and the descriptions of most were collected in a general report by Dr. Girard. The main features of the ichthyology of the Pacific slope were then already known; but more recent laborers have not only extended largely our knowledge of species, but added a number of entirely new forms, and thrown much light on the relations of the fish-fauna of that region to others.

The following bibliography is a nearly complete enumeration, in chronological order, of the memoirs and articles of all kinds that have been published on the fishes of the region in question. The chronological order has been determined by the date of reading of the articles

J

communicated to learned societies. In cases of question of priority, the right depends, of course, on the period of publication; but this is sometimes with great difficulty ascertainable, and motives of convenience have dictated the sequence adopted.

Perhaps some will be disposed to believe that the compiler has sinned in redundancy rather than deficiency in this bibliography. The evils of the former are, however, easily remedied, while those of the latter must leave the consulter in more or less doubt. Many popular works have been catalogued where original information of even slight value was contained, and when such works were among the earliest published on the regions.in question. Besides those enumerated, works on California, too numerous to mention, contain incidental information (very rarely of any original value, however) respecting the fishes and fisheries of that State; and a number on the British possessions belong to the same category. Among those relative to British Columbia and Vancouver's Island worthy to be mentioned, but not to be particularized, are the volumes of Wm. Carew Hazlitt (1858), J. Desford Pemberton (1860), Duncan George Forbes Macdonald (1862), Capt. C. E. Barrett Lennard (1862), Alexander Rattray (1862), Com. R. C. Mayne (1862), G. M. Sproat (1868), Francis Poole (1872), and Capt. W. F. Butler (1873).

The titles of the Government publications are taken from a manuscript compilation embracing notices of all the reports published by the General and State governments on scientific explorations, and intended to be more particular than the present work. They are retained with the bars (|), indicating the distribution on the title-pages of the lines, etc.

Several societies have, or have had, the custom of publishing communications, sometimes of an elaborate and extended nature, without any titles. This strange and senseless mode of procedure seems to have originated in some freak or affectation of modesty on the part of authors, perhaps, rather than a deliberate intention to shirk labor or confuse matters. Confusion and trouble to others are nevertheless the result of this vicious negligence, and a consequence is an ignoring of the papers thus unentitled or an irreconcilable variation of titles in different bibliographies. Whether the custom originates with authors or not, the assumption of it is discreditable to the editor or editors of the publications adopting it. A number of the papers here recorded belong to this category of the unentitled or disentitled: the titles following preceded by an asterisk (*) are selected from the remarks prefatory to the paper in the proceedings, and those preceded by a dagger (†) have been composed by the present writer, since nothing intelligible precedes the papers themselves. It is to be hoped that the senseless and causeless sin in question may speedily be discontinued. There is no reason why any one should be compelled to read the whole of an article (as is sometimes necessary) to obtain an idea of what the paper relates to; and the "Catalogue of Scientific Papers (1800–1863) compiled and published by the Royal Society of London" shows how a bibliography edited under the best auspices may be involved in grave errors by the negligence adverted to.

TITLES OF WORKS.

1757—Noticia de la California, y de su·conquista temporal y espiritual hasta el tiempo presente. Sacada de la historia manuscripta, formada en Mexico año de 1739. por el Padre **Miguel Venegas**, de la Compañia de Jesus; y de otras Noticias, y Relaciones antiguas, y modernas. Añadida de algunos mapas particulares, y uno general de la America Septentrional, Asia Oriental, y Mar del Sùr intermedio, formados sobre las Memorias mas recientes, y exactas, que se publican juntamente. Dedicada al Rey N.ᵗᵉᵒ Señor por la Provincia de Nueva-España, de la Compañia de Jesus. Tomo primero [—Tomo tercero].—Con licencia. En Madrid : En la Imprenta de la Viuda de Manuel Fernandez, y del Supremo Consejo de la Inquisicion. Año de M.D.CCLVII. [8°, 3 vols.]

[Translated as follows :—]

A Natural and Civil History of California: containing an accurate description of that country, its soil, mountains, harbours, lakes, rivers, and seas ; its animals, vegetables, minerals, and famous fishery for pearls. The customs of the inhabitants, their religion, government, and manner of living, before their conversion to the Christian religion by the missionary Jesuits. Together with accounts of the several voyages and attempts made for settling California, and taking actual surveys of that country, its gulf, and coast of the South-Sea. Illustrated with copperplates, and an accurate map of the country and adjacent seas. Translated from the original Spanish of **Miguel Venegas**, a Mexican Jesuit, published at Madrid 1758.—In two volumes.—Vol. I[—II]. = London : printed for James Rivington and James Fletcher, at the Oxford Theatre, in Pater-Noster-Row. 1759. [8°, vol. i, 10 l., 455 pp., 1 pl. ; vol. ii.]

[The only references to fishes are as follows (v. i, pp. 47-48) :—"But if the soil of California be in general barren, the scarcity of provisions is supplied by the adjacent sea ; for both in the Pacifick ocean and the Gulf of California, the multitude and variety of fishes are incredible. Father Antonio de la Ascencion, speaking of the bay of San Lucas [Lower California], says, ' With the nets which every ship carried, they caught a great quantity of fish of different kinds, and all wholesome and palatable: particularly holybuss, salmon, turbots, skates, pilchards, large oysters, thornbacks, mackerel, barbels, bonetos, soals, lobsters, and pearl oysters.' And, speaking of the bay of San Francisco, on the western coast, he adds : ' Here are such multitudes of fish, that with a net, which the commodore had on board, more was caught every day than the ship's company could make use of : and of these a great variety, as crabs, oysters, breams, mackerel, cod, barbels, thornbacks, &c.' And in other parts he makes mention of the infinite number of sardines, which are left on the sand at the ebb; and so exquisite that those of Laredo in Spain, then famous for this fish, do not exceed them. Nor are fish less plentiful along the gulf [of California], where to the above mentioned species Father Picolo adds, tunnies, anchovies, and others. Even in the rivulets of this peninsula are found barbels and crayfish : but the most distinguished fish of both seas are the whales ; which induced tho· ancient cosmographers to call California, Punta de Balenas, or Cape Whale : and these fish being found in multitudes along both coasts, give name to a channel in the gulf, and a bay in the South sea" (v. i, pp. 47-48).]

7

1772—Voyage en Californie pour l'observation du passage de Vénus sur le disque du soleil, le 3 juin 1769; contenant les observations de ce phénomène et la discription historique de la route de l'auteur à travers le Mexique. Par feu M. **Chappe d'Auteroche**, . . . Rédigé et publié par M. de Cassini fils . . . À Paris: chez Charles-Antoine Jombert. MDCCLXXII. [4°, half-title, title, 170 [2] pp., plan, and 2 pl.—Sabin.]

[Translated as follows:—]

A Voyage to California, to observe the Transit of Venus. By Mons. **Chappe d'Auteroche**. With an historical description of the author's route through Mexico, and the natural history of that province. Also, a voyage to Newfoundland and Sallee, to make experiments on Mr. Le Roy's time keepers. By Monsieur de Cassini. London: printed for Edward and Charles Dilly, In The Poultry. MDCCLXXVIII. [8°, 4 p. l., 315 pp., with "plan of City of Mexico".]

> Extract of a letter from Mexico addressed to the Royal Academy of Sciences at Paris, by Don **Joseph Anthony de Alzate y Ramyrez**, now a correspondent of the said academy, containing some curious particulars relative to the natural history of the country adjacent to the City of Mexico. pp. 77–105.
>
> [It is undoubtedly this work that is meant in the statement that has so largely gone the rounds of the periodical press, to the effect that the Californian viviparous fishes were observed during the voyage for the observation of the transit of Venus to Lower California, 1769. A perusal of the accounts given, however, renders it evident that the fishes in question were not Embiotocida but rather Cyprinodontida, probably of the genus *Mollienesia*. The account by Don Alzate (pp. 89-91) is as follows:—
>
> "I send you some viviparous scaly fishes, of which I had formerly given you an account. What I have observed in them this year is—'If you press the belly with your fingers, you force out the fry before their time, and upon inspecting them through the microscope you may discern the circulation of the blood, such as it is to be when the fish is grown up.' If you throw these little fishes into water, they will swim as well as if they had been long accustomed to live in that element. The fins and tail of the males are larger and blacker than those of the females, so that the sex is easily distinguished at first sight. These fish have a singular manner of swimming; the male and the female swim together on two parallel lines, the female always uppermost and the male undermost; they thus always keep at a constant uniform distance from each other, and preserve a perfect parallelism. The female never makes the least motion, either sideways or towards the bottom, but directly the male does the same."
>
> To this account is added a foot-note (p. 90) containing the following additional information:—
>
> "Don Alzate has sent those fishes preserved in spirits; their skin is covered with very small scales; they vary in length from an inch to eighteen lines, and they are seldom above five, six, or seven lines in the broadest part. They have a fin on each side near the gills, two small ones under the belly, a single one behind the anus, which lies between the fin and the single one; the tail is not forked; lastly, this fish has a long fin on the back, a little above the fin, which is under the belly.
>
> "We know of some viviparous fishes in our seas, such as loach, &c. most of these have a smooth skin without any scales. The needle of Aristotle is viviparous, and yet covered with broad and hard scales, I have caught some that had young ones still in their womb. As to these viviparous fishes, it is a particular and new sort, and we are obliged to Don Alzate for making us acquainted with it. It breeds in a lake of fresh water near the City of Mexico."
>
> This is, so far as known, the earliest notice of the viviparity of Cyprinodontida. The mode of consorting together (exaggerated in the account) is common to a number of representatives of the family, and is alluded to by Prof. Agassiz in a name (*Zygonectes, i. e.* swimming in pairs) conferred on one of the genera of the family.]

1808—Piscium Camtschaticorum [*Terpuk*] et [*Wachnja*]. Descriptiones et icones auctore [**W. G.**] Tilesio. D. 26 Octobri 1808. Conventui exhib. die 2 Nov. 1808. < Mém. Acad. Sci. Pétersb., v. 2, pp. 335–375, 1810, viz:—

> I. Hexagrammos Stelleri, Rossis Terpuc dictus novum genus piscium Camtschaticorum. pp. 335–340, tab. 15.

II. Dimensiones piscis, beato Stellero Hexagrammos asper dicti, Rossis Teerpuk [*Terpuk*] i. e. lima (captns d. 20 Maij 1741 in portu Divi Petri et Pauli pondebat pondere medicinali duas usque ad sex uncias). pp. 340-341.

III. Hexagrammos Stelleri, quænam genera sit interponendus cuinam classi ordinique systematico sit inserendns. Labrax Pallasii (vid. ej. Monograph.). pp. 342-343.

IV. Descriptio Stelleri anno 1741 concepta. pp. 343-347.

V. Observationes anatomicæ. pp. 347-349.

VI. Wachnja Camtschatica est Gadus dorso tripterygio, Callariis speciatim Lusco affinis. pp. 350-353, tab. 16, 17.

VII. Wachniæ Camtschaticæ altera species, (Gadus gracilis mihi,) quæ ab indigenis Camtschaticis acque Üachal, Rossis Wachnja [*Wachnja*] dicitur, dimensionibus illustrata. pp. 354-356, tab. 18.

VIII. Stelleri Descriptio piscis ονος sive asini antiquorum. Turneri ad Gesnerum aselli 3 sivi Æglefini Rondelet et Gesneri. Æglefini Bellonii, Anglorum Hadok, Russis Wachnja [*Wachnja*] dicti corrupta voce Itaelmannica, in qua Üakal audit. pp. 356-359.

IX. Observationes anatomicæ. pp. 360-363.

X. Observationes ex aliorum individuorum ejusdem speciei dissectionibus, pp. 363-364.

XI. Ad historiam Gadi dorso tripterygio ore cirrato caudo æquali fere cum radio primo spinoso (Kabeljau vel Cabiljan Belgarum) (Gadus morrhua L. Bloch. tab. 64), adhuc annotata sequentia. pp. 364-370.

XII. Annotationes anatomicæ. pp. 370-371.

XIII. Tabularum explicatio. pp. 372-375.

1809—Labraces, novum genus piscium, oceani orientalis, auctore **P. S. Pallas.** Conventui exhib. die 5 Julii 1809. < Mém. Acad. Sci. St. Pétersb., v. 2, pp. 382-398, 1810.

[N. sp. *L. decagrammus, L. superciliosus, L. monopterygius.*]

Description de quelques poissons observés pendant son voyage antour du monde. Par **W. G. Tilesius.** < Mém. Soc. Imp. des Naturalistes de Moscou. t. 2, pp. 212-249, with 5 pl., 1809.

1811—Iconum et Descriptionum piscium Camtschaticorum continuatio tertia tentamen monographiæ generis Agoni Blochiani sistens. Auctore [**W. G.**] Tilesio. Cum tabulis vi æneis.—Conventui exhibita die 11 Decembris 1811. < Mém. Acad. Sci. Pétersb., v. 4, pp. 406-478, 1813, viz:—

De novis piscium generibus, Agono Blochii et Phalangiste cel. Pallasii, propter synonymiam conjugendis. pp. 406-454.

Appendix de Cyprino rostrato et cultrato, Trachino trichodonto et Epenephelo ciliato. pp. 454-457.

Descriptio Cyprini rostrati Tungusis ad Covymam fluv. Tschukutscham et Jucagiris Ouatscha dicti. pp. 457-474, tab. xv, fig. 1-5.

Epinephelus ciliatus Camtschaticus et Americanus. pp. 474-478, tab. xvi, fig. 1-6.

Zoographia Rosso-Asiatica, sistens Omnium Animalium in extenso imperio Rossico et adjacentibus maribus observatorum Recensionem, Domicilia, Mores et Descriptiones, anatomen atque Icones plurimorum. Auctore

Petro Pallas, Eq. Aur. Academico-Petropolitano.—Volumen tertium.—Petropoli in Officina Caes. Academiæ Scientiarum Impress. M.DCC.CXI. Edit. MDCCCXXXI. [4°, vii, 428, cxxv pp., 6 pl.]

[As indicated on the title-page, the "Zoographia Rosso-Asiatica" was not regularly *published* till 1831, but was printed in 1811, and was only detained by the loss of the copper-plates. The letter-press was, however, to a slight extent, distributed before the regular publication of the edition, and a copy was possessed by Cuvier, who has given a summary of the third volume in the Histoire Naturelle des Poissons (t. 1, pp. 200–201).

Describes species of which specimen had been obtained from the Russian possessions in Northwestern America. The following are published as if new, although several had previously been described:—

    Phalangistes acipenserinus (p. 110, pl. 17).
    Cottus polyacanthocephalus (p. 133, pl. 23).
    Cottus platycephalus (p. 135, pl. 24).
    Cottus trachurus (p. 138, pl. 25).
    Cottus pistilliger (p. 143, pl. 20, f. 3, 4).
    Blennius dolichogaster (p. 173, pl. 42, f. 2).
    Blennius anguillaris (p. 176, pl. 42, f. 3).
    Gadus wachna (p. 182, pl. 44).
    Gadus pygmœus (p. 199).
    Gadus fimbria (p. 200).
    Ammodytes hexapterus (p. 226).
    Ammodytes septipinnis (p. 227, pl. 48, f. 3).
    Trachinus trichodon (p. 235, pl. 50, f. 1).
    Trachinus cirrhosus (p. 237, pl. 50, f. 2).
    Perca variabilis (p. 241).
    Labrax decagrammus (p. 278, pl. 62, f. 2).
    Labrax superciliosus (p. 279, pl. 63, f. 1).
    Labrax monopterygius (p. 281, pl. 63, f. 4).
    Labrax octogrammus (p. 283, pl. 64, f. 1).
    Salmo lagocephalus (p. 372, pl. 77, f. 2).
    Salmo proteus (p. 376, pl. 78, f. 2, pl. 79).
    Pleuronectes quadrituberculatus (p. 423).
    Pleuronectes cicatricosus (p. 424).

The plates referred to were never published.

The only other species signalized as inhabitants of the American waters are the following:—

    Raja batis (p. 57).
    Salmo socialis (p. 389, pl. 81, f. 2).
    Pleuronectes stellatus (p. 416).
    Pleuronectes hippoglossus (p. 421).]

1814—History | of | the expedition | under the command of | Captains Lewis and Clark, | to | the sources of the Missouri, | thence | across the Rocky Mountains | and down the | River Columbia to the Pacific Ocean. | Performed during the years 1804-5-6. | By order of the | Government of the United States. | Prepared for the press | by **Paul Allen**, Esquire. | In two volumes. | Vol. I [—II]. | Philadelphia: | Published by Bradford and Inskeep; and | Abm. H. Inskeep, Newyork. | J. Maxwell, Printer. | 1814. [8°, vol. i, lxxviii, 470 pp., maps; vol. ii, ix, 522 pp., maps.]

[Vol. ii, chap. vii, contains "A general description of the beasts, birds, and plants, &c., found by the party in this expedition" (pp. 148–201). Incidental allusions and quasi-descriptions of a popular kind are given of some fishes, but nothing of an exact nature is made known.

"An account of the various publications relating to the travels of Lewis and Clarke, with a commentary on the zoological results of their expedition", has been published by Dr. Elliott Coues, U. S. A. (Bull. U.! S. Geol. and Geog. Surv. Terr., v. 1, pp. 417–444, Feb. 8, 1876).]

1820—Relation d'un voyage à la côte du nord-ouest de l'Amérique septentrionale dans les années 1810-1814. Par **Gabriel Franchère**. [Rédigé par Michel Bibaud.] Montréal, 1820. [8°, 284 pp.—Sabin.]

[Translated as follows:—]

Narrative of a voyage to the northwest coast of America in the years 1811, 1812, 1813, and 1814, | or the first American settlement on the Pacific | By Gabriel Franchere | Translated and edited by J. V. Huntington | — | Redfield | 110 and 112 Nassau street, New York | 1854. [12°, 376 pp., 3 pl.]

[The salmon is noticed in chapter 18.]

1822—Voyage pittoresque autour du monde, avec des portraits de sauvages d'Amérique, d'Asie, d'Afrique, et des îles du grand océan ; des paysages, des vues maritimes, et plusieurs objets d'histoire naturelle; accompagné de descriptions par M. le Baron Cuvier, et M. A. de Chamisso, et d'observations sur les crânes humains par M. le Docteur Gall. Par M. **Louis Choris**, Peintre.— Paris, de l'imprimerie de Firmin Didot, . . . 1822. [Fol., 2 p. l., vi pp.+[i], 12 pl., 17 pp.+[ii], 10 pl., 20 pp.+[iii], 14 pl., 10, 3 pp.+[iv], 18 pl., 24 pp.+ [v], 19 pl., 22 pp. + [vi], 23 pl., 28 pp. + [vii], 7 pl., 19 pp.]

[Partie vi.] Chapeau de bois, sur lequel sont peintes divers animaux marins. Planche v. Par G. Cuvier. pp. 21-22.

[Cuvier considers that one of the figures (*h*) represents a *Diodon*, and such *seems* to be the case; but no species of that type has been found so far northward as Unalashka, where the hat was obtained. ("En *h*, est un *Diodon* ou orbe épineux, qui est pris à la ligne tandis que les grands cétacés du reste de ce tableau sont poursuivis avec des lances" (p. 22).]

1823—Account | of | an expedition | from | Pittsburgh to the Rocky Mountains, | performed in the years 1819 and '20, | by order of | the Hon. J. C. Calhoun, Sec'y of War: | under the command of | Major Stephen H. Long. | From the notes of Major Long, Mr. T. Say, and other gen- | tlemen of the exploring party. | — | Compiled | by **Edwin James**, | botanist and geologist for the expedition. | — | In two vols.—With an atlas. | Vol. II. | — | Philadelphia: | H. C. Carey and J. Lea, Chesnut st. | 1823. [2 v., 8°. Vol. i, 2 p. l., 503 pp.; vol. ii, 3 p. l., 442 pp.]

1828—Histoire Naturelle des Poissons, par M. le Bᵒⁿ **Cuvier**, . . . ; et par M. **Valenciennes**, ... Tome premier. À Paris, chez F. G. Levrault, ... 1828. [8° ed. xvi, 574 pp., 1 l.; 4° ed. xiv, 422 pp., 1 l.—pl. 1-8 (double).]
Livre premièr.—Tableau historique des progrès de l'ichthyologie, depuis son origine jusqu'à nos jours.
Livre deuxième.—Idée générale de la nature et de l'organisation des poissons.

[Pallas' "Zoographia Rosso-Asiatica" noticed at pp. 200-201.]

Histoire Naturelle des Poissons, par M. le Bᵒⁿ **Cuvier**, . . . ; et par M. **Valenciennes**, ... Tome deuxième. À Paris, chez F. G. Levrault, . . . 1828. [8° ed. xxi, (1 l.), 490 pp.; 4° ed. xvii, (1 l.), 371 pp.—pl. 9-40.]
Livre troisième.—Des poissons de la famille des Perches, ou des Percoïdes. [Par Cuvier.]

[No west-coast species specified.]

1829—Histoire Naturelle des Poissons, par M. le Bᵒⁿ **Cuvier**, . . . ; et par M. **Valenciennes**, ... Tome troisième. À Paris, chez F. G. Levrault, . . . , 1829. [8° ed. xxviii, 500 pp., 1 l.; 4° ed. xxii, (1 l.), 368 pp.—pl. 41-71.]
Livre troisième.—Des poissons de la famille des Perches, ou des Percoïdes. [Par Cuvier.]

[N. sp. name, *Trichodon Stelleri*, based on *Trachinus trichodon* Pallas.]

1829—Histoire Naturelle des Poissons, par M. le B⁰ⁿ **Cuvier**, . . . ; et par M. **Valen-**
ciennes, ... Tome quatrième. À Paris, chez F. G. Levrault, . . . ,
1829. [8° ed xxvi, (1 1.), 518 pp.; 4° ed. **xx,** (1 1.), 379 pp.—pl. 72-99, 97 bis.
Livre quatrième.—Des Acanthoptérygiens à joue cuirassée. [Par Cuvier.]
[N. sp. *Cottus ventralis, Hemilepidotus Tilesii.*]

Zoologischer Atlas, enthaltend Abbildungen und Beschreibungen neuer Thier-
arten, während des Flottcapitains von Kotzebue zweiter Reise um die
Welt, auf der Russisch-Kaiserlichen Kriegsschlupp Predpriatië in den
Jahren 1823-1826 beobachtet von Dr. **Friedr. Eschscholtz,** Professor und
Director des zoologischen Museums an der Universität zu Dorpat, Mitglied
mehrerer gelehrten Gesellschaften, Russ. Kais. Hofrathe und Ritter des
Ordens des heil. Wladimir. Drittes Heft.—Berlin, 1829. Gedruckt und
verlegt bei G. Reimer. [Fol., title, 18 pp., pl. 11-15.]
[N. sp. *Blepsias ventricosus* (p. 4, pl. 13), on which was subsequently based the genus
Temnistia of Richardson.]

1830—Histoire Naturelle des Poissons, par M. le B⁰ⁿ **Cuvier**, . . . ; et par M. **Valen-**
ciennes, ... Tome cinquième. À Paris, chez F. G. Levrault, . . . ,
1830. [8° ed. xviii, 499 pp., 2 1.; 4° ed. **xx,** 374 pp., 2 1.—pl. 100-140.]
Livre cinquième.—Des Sciénoïdes. [Par Cuvier.]
[No west-coast species noticed.]

Histoire Naturelle des Poissons, par M. le B⁰ⁿ **Cuvier**, . . . ; et par M. **Valen-**
ciennes, ... Tome sixième. À Paris, chez F. G. Levrault, . . . ,
1830. [8° ed. xxiv, 559 pp., 3 1.; 4° ed. xviii, (3 1.), 470 pp.—pl. 141-169. 162
bis, 162 ter, 162 quater, 167 bis, 168 bis.]
Livre sixième.—(Partie I.—Des Sparoïdes. Partie II.—Des Ménides.)
[Par Cuvier et Valenciennes.]
[No west-coast species noticed.]

1831—Histoire Naturelle des Poissons, par M. le B⁰ⁿ **Cuvier**, . . . ; et par M. **Valen-**
ciennes, ... Tome septième. À Paris, chez F. G. Levrault, . . . ,
1831. [8° ed. **xxix,** 531 pp., 3 1.; 4° ed. xxii, (3 1.), 399 pp.—pl. 170-208.]
Livre septième.—Des Squamipennes. [Par Cuvier ?]
Livre huitième.—Des poissons à pharyngiens labyrinthiformes. [Par
Cuvier ?]
[No west-coast species noticed.]

Histoire Naturelle des Poissons, par M. le B⁰ⁿ **Cuvier**, . . . ; et par M. **Valen-**
ciennes, ... Tome huitième. À Paris, chez F. G. Levrault, . . . ,
1831. [8° ed. xix, (2 1.), 509 pp.; 4° ed. xv, (2 1.), 375 pp.—pl. 209-245.]
Livre neuvième.—Des Scombéroïdes. [Par Cuvier et Valenciennes.]
[No west-coast species noticed.]

Zoographia Rosso-Asiatica. See 1811.

1833—Histoire Naturelle des Poissons, par M. le B⁰ⁿ **Cuvier**, . . . ; et par M. **Valen-**
ciennes, ... Tome neuvième. À Paris, chez F. G. Levrault, . . . ,
1833. [8° ed. xxix, 512 pp., 1 1.; 4° ed. xxiv, (1 1.), 379 pp.—pl. 246-279.]
Livre neuvième.—Des Scombéroïdes. [Par Cuvier et Valenciennes.]
[No west-coast species noticed.]

1835—Histoire Naturelle des Poissons, par M. le Bon Cuvier, . . . ; et par M. Valenciennes, ... Tome dixième. À Paris, chez F. G. Levrault, . . . , 1835. [8° ed. xxiv, 482 pp., 1 l.; 4° ed. xix, (1 l.), 358 pp.—pl. 280–306.]
Suite du livre neuvième—Des Scombéroïdes. [Par Cuvier et Valenciennes ?]
Livre dixième.—De la famille des Teuthies. [Par Cuvier et Valenciennes ?]
Livre onzième.—De la famille des Tænioïdes. [Par Cuvier et Valenciennes ?]
Livre douzième.—Des Atherines. [Par Cuvier et Valenciennes ?]
[No west-coast species noticed.]

1836—Fauna Boreali-Americana; or the Zoology of the Northern Parts of British America: containing descriptions of the objects of Natural History collected on the late northern land expeditions under command of Captain Sir John Franklin, R. N. Part third. The Fish. By John Richardson, M. D., F. R. S., F. L. S., Member of the Geographical Society of London, and Wernerian Natural History Society of Edinburgh; Honorary Member of the Natural History Society of Montreal, and Literary and Philosophical Society of Quebec; Foreign Member of the Geographical Society of Paris; and Corresponding member of the Academy of Natural Sciences of Philadelphia; Surgeon and Naturalist to the Expeditions.—Illustrated by numerous plates.—Published under the authority of the Right Honourable the Secretary of State for Colonial Affairs.—London: Richard Bentley, New Burlington street, MDCCCXXXVI. [4°, pp. xv, 327 (+1) pp., 24 pl. (numbered 74–97).]

[N. g. and n. sp. *Temnistia* (n. g., 59), *Cyprinus* (*Leuciscus*) *gracilis* (120), *Salmo Scouleri* (158, 223), *Salmo quinnat* (219), *Salmo Gairdneri* (221), *Salmo paucidens* (222), *Salmo tsuppitch* (224), *Salmo Clarkii* (225, 307), *Salmo* (*Mallotus?*) *pacificus* (226), *Acipenser transmontanus* (278), *Petromyzon tridentatus* (293); (ADDENDA:) *Cottus asper* (295, 313), *Cyprinus* (*Abramis*) *balteatus* (301), *Cyprinus* (*Leuciscus*) *caurinus* (304), *Cyprinus* (*Leuciscus*) *oregonensis* (305).]

Report on North American Zoology. By John Richardson, M. D., F. R. S. < Rep. 6th meeting Brit. Assoc. Adv. Sci., Aug. 1836, = v. 5, pp. 121–224, 1837.

Pisces, pp. 202–223.

Astoria, or anecdotes of an enterprise beyond the Rocky Mountains. By Washington Irving. [1st ed.] In two volumes. Vol. I [—II]. Philadelphia: Carey, Lea & Blanchard. 1836. [2 vols., 8°. Vol. i, 285 pp.; vol. ii, 279 pp., 1 map folded.]
[The fishes and fisheries, especially salmon, are noticed in vol. 2, chapters 9 and 14.]

Histoire Naturelle des Poissons, par M. le Bon Cuvier, ...; et par M. Valenciennes, ... Tome onzième. À Paris, chez F. G. Levrault, . . . , 1836. [8° ed. xx, 506 pp., 1 l.; 4° ed. xv, (1 l.), 373 pp.—pl. 307–343.]
Livre troisième.—Des Mugiloïdes.
Livre quatorzième.—De la famille des Gobioïdes.
[No west-coast species noticed.]

1837—Histoire Naturelle des Poissons, par M. le Bon Cuvier, . . . ; et par M. Valenciennes, . . . Tome douzième. À Paris, chez F. G. Levrault, . . . , 1837. [8° ed. xxiv, 507 +1 pp.; 4° ed. xx, 377 pp., 1 l.—pl. 344–368.]
Suite du livre quatorzième.—Gobioïdes.
Livre quinzième.—Des Acanthoptérygiens à pectorales pédiculées.

1839—Histoire Naturelle des Poissons, par M. le Bon Cuvier, . . . ; et par M. Valenciennes, . . . Tome troisième. À Paris, chez Pitois-Levrault et Ce, . . . , 1839. [8° ed. xix, 505 pp., 1 l.: 4° ed. xvii, 370 pp.—pl. 369–388.]
Livre seizième—Labroïdes.
[No west-coast species noticed.]

1839—Histoire Naturelle des Poissons, par M. le Bon **Cuvier**,...; et par M. **Valenciennes**, ... Tome quatorzième. À Paris, chez Pitois-Levrault et Co,...,
1839. [8° ed. xxii, 464 pp., 3 l.; 4° ed. xx, 344 pp., 3 l.—pl. 389–420.]
Suite du livre seizième.—Labroïdes.
Livre dix-septième.—Des Malacoptérygiens. Des Siluroïdes.

[No west-coast species noticed.]

The Zoology of Captain Beechey's Voyage; compiled from the collections and notes made by Captain Beechey, the officers and naturalist of the Expedition, during a Voyage to the Pacific and Behring's straits performed in his Majesty's Ship Blossom, under the command of Captain F. W. Beechey, R. N., F. R. S., &c., &c. in the years 1825, 26, 27, and 28. By J. Richardson, M. D., F. R. S., &c.; N. A. Vigors, Esq., A. M., F. R. S., &c.; G. T. Lay, Esq.; E. T. Bennett, Esq., F. L. S., &c.; the Rev. W. Buckland, D. D., F. R. S., F. L. S., F. G. S., &c. and G. B. Sowerby, Esq.—Illustrated with upwards of fifty finely coloured plates, by Sowerby.—Published under the authority of the Lords Commissioners of the Admiralty. = London : Henry G. Bohn, 4, York Street, Covent Garden.—MDCCCXXXIX.

    Fishes; by **G. T. Lay**, Esq., and **E. T. Bennett**, Esq., F. L. S., &c. pp. 41–75, pl. 15–23.

[N. sp. *Ohimœra colliei* (p. 71, pl. 23).

This volume is interesting as being the first publication in which any attempt has been made to scientifically indicate the fishes of the coast. The "naturalist" of the expedition was, however, incompetent for the task, and the notes taken evince that he was not sufficiently versed in the rudiments of ichthyology to know what to observe. Nevertheless, the notes have an interest, if not of importance, enough to transcribe what relates to the regions in question:—

"Off Saint Lawrence Island was caught, in the dredge a fish apparently allied to the genus *Liparis*, Art. It had the 'ventral fins placed before the pectorals, but united and continuous with them; a flat, raised, and rough tubercle, of nearly the diameter of an English sixpence, was seated forward between the pectorals, its anterior part reaching as far as the ventrals; this may be of use in copulation : its *cœca* were pretty numerous.'—C. The roughness of this tubercle renders it difficult to refer the fish to any known species; but it is probably nearly related to the *Cyclopterus gelatinosus*, Pall., a *Liparis* which is known to inhabit the seas in which this was obtained. The existence of *cœca* removes it from *Lepadogaster*, Gouan.

"Kotzebue Sound afforded a specimen of a new species of *Ophidium*, L., the *Oph. stigma*.

"On the coast of California, a little to the northwards of the harbour of San Francisco, an *Orthagoriscus* was met with, apparently the *Orth. mola.*, Bl. They swam about the ship with the dorsal fin frequently elevated above the surface." (p. 50.)

"On the coast of California, at Monterey, Mr. Collie's notes mention the occurrence of [1] a species of *Sparus*, of two *Scombri*, and of a *Clupea*. [2] The first of the *Scombridœ* is apparently a *Scomber*, Cuv.; it was 'smaller than the mackerel; it was marked on the back with cross waved narrow bands of black and greenish blue; its first dorsal fin had nine spines, and there were four small pinnules behind the second dorsal and the anal: it had a simple air-bladder of moderate size, and an immense number of *cœca*, with a stomach extending the whole length of the abdomen, narrow, tapering to the posterior part, and covered throughout nearly its whole length with the milt.? Its internal membrane forms longitudinal folds; the intestines have three convolutions.'—C. This fish occurred in shoals. [3] The second species was met with but once. It is a *Caranx*, Cuv., of which 'the teeth in the upper maxillary are scarcely to be felt: the pectorals reach nearly to opposite the *anus:* a double narrow stripe of deeper blue than the general surface runs backwards on each side of the first dorsal fin to opposite its termination, the two parts being separated by a broad line of dirty white, which has a narrow, dark-coloured line along its middle: there are no distinct divisions in the anal and second dorsal fins: the air-bladder is simple, and small, and extends from the *fauces* to the *anus;* the stomach is much shorter than in the preceding species; the *cœca*, although numerous, are less so than in it, and the intestine is folded in the same manner.'—C. From the nature of the colouring of this fish, as described by Mr. Collie, there can be little doubt of its constituting a distinct species.

[4] Along with the first species of *Scomber*, there occurred in shoals a small species of *Clupea*, L., 'without teeth; with the dorsal fin a little before the ventral; and with the back dark greenish blue, and having one line and part of another of rounded black spots on each side nearly on a level with the eye: the gill membranes contain six rays, and overlap each other at their lower part; the stomach resembles that of the first *Scomber*; it has also numerous *cœca*; the air-bladder is small and tapering.'—C. The other fishes observed at Monterey were [5] a new species of *Ohimœra*, Cuv., differing essentially from the *Chimœra* of the Atlantic, and approaching somewhat in the position of its second dorsal fin to the *Callorhynchus*, Cuv.; [6] a species of *Torpedo*, Dum.; and [7] a *Raia* " (pp. 54–55).]

1839—Narrative of a Journey across the Rocky Mountains, to the Columbia River, and a Visit to the Sandwich Islands, Chili, &c. With a Scientific Appendix. By John K. **Townsend**, Member of the Academy of Natural Sciences of Philadelphia. Philadelphia: Henry Perkins, 134 Chestnut street. Boston: Perkins & Marvin.—1839. [8°, 352 pp.]

[A few incidental popular notices of salmon and trout are given.]

[Reprinted in England under the following title:—]

Sporting Excursions in the Rocky Mountains, including a Journey to the Columbia River, and a Visit to the Sandwich Islands, Chili, &c. By J. K. **Towshend** [*sic!*], Esq. In two volumes. Vol. I [—II]. London: Henry Colburn, Publisher, Great Marlborough Street. 1840. [8°. Vol. i, xii [+i], 312 pp., 1 pl.; vol. ii, xii, 310 pp., 1 pl.]

[In vol. i, chap. 7, are given details respecting salmon and the mode of catching them, and the frontispiece illustrates a native woman "spearing the salmon".]

1840—Histoire Naturelle des Poissons, par M. le Bᵒⁿ **Cuvier**, . . .; et par M. **Valenciennes**, . . . Tome quinzième. À Paris, chez Ch. Pitois, éditeur, . . . , 1840. [8° ed. xxxi, 540 pp., 1 l.; 4° ed. xxiv, 397 pp.—pl. 421–435.]

Suite du livre dix-septième.—Siluroïdes.

[No west-coast species noticed.]

Narrative of a whaling voyage round the globe, from the year 1833 to 1836, comprising sketches of Polynesia, California, the Indian Archipelago, etc. with an account of Southern Whales, the Sperm Whale Fishery, and the Natural History of the climates visited. By **Frederick Debell Bennett**, Esq., F. R. G. S., Fellow of the Royal College of Surgeons, London. In two volumes. Vol. I [—II]. London: Richard Bentley, New Burlington street, publisher in ordinary to her Majesty.—1840. [8°, vol. i, xv, 402 pp., 1 pl., 1 map; vol. ii, vii, 396 pp., 1 pl.]

1842—Histoire Naturelle des Poissons, par M. le Bᵒⁿ **Cuvier**, . . . , et par M. **Valenciennes**, ... Tome seizième. À Paris, chez P. Bertrand, . . . , 1842. [9° ed. xx, 472 pp., 1 l.; 4° ed. xviii, 363 pp., 1 l.—pl. 456–487.]

Livre dix-huitième.—Cyprinoïdes.

Zoology of New-York, or the New-York Fauna; comprising detailed descriptions of all the animals hitherto observed within the State of New-York, with brief notices of those occasionally found near its borders, and accompanied by appropriate illustrations.—By James E. **DeKay**.—Part IV.—Fishes. Albany: Printed by W. & A. White and I. Visscher. 1842. [4°, xiv [1, errata], 415 pp.; atlas, 1 p. l., 79 pl.]

[The letterpress of the Reptiles and Fishes, each separately paged, forms one volume, and the plates, each separately numbered, another. Eight of the northwest-coast Malacopterygian species (*Abramis balteatus, Leuciscus caurinus, Leuciscus oregonensis, Salmo quinnat, Salmo Gairdnerii, Salmo Scouleri, Salmo tsuppitch,* and *Salmo nitidus*) and the Sturgeon (*Acipenser transmontanus*) enumerated by Richardson (1836) are briefly indicated as "extra-limital".]

1844—Histoire Naturelle des Poissons, par M. le Bon **Cuvier**, . . .; et par M. Valenciennes, ... Tome dix-septième. À Paris, chez P. Bertrand, . . . , 1844. [8° ed. xxiii, 497 pp., 1 l.; 4° ed. xx, 370 pp. 1 l.—pl. 487 (bis)–519.] Suite du livre dix-huitième.—Cyprinoïdes.

1845—Description of a new species of *Syngnathus*, brought from the western coast of California by Capt. Phelps. By Dr. **D. H. Storer**. < Proc. Boston Soc. Nat. Hist., v. 2, p. 73, December, 1845.

[N. sp. *Syngnathus californiensis.*]

1846—A Synopsis of the Fishes of North America. By **David Humphreys Storer**, M. D., A. A. S., ... < Mem. Am. Acad. Arts and Sci., new series, vol. ii, pp. 253–550, Cambridge, 1846.

[739 nominal species from all North America, including the West Indies, are described. The descriptions, however, are most inaptly compiled and entirely insufficient.]

A Synopsis of the Fishes of North America. By **David Humphreys Storer**, M. D., A. A. S., ... Cambridge: Metcalf and Company, Printers to the University. 1846. [4°, 1 p. l. (= title), 298 pp.]

[A reprint, with separate pagination, title-page, and index, of the preceding.

According to Dr. Storer (Mem. Acad., p. 260; Syn. p. 8), "the following species inhabit the northwestern coast of America:—

Trichodon stelleri.	*Salmo salar.*
Cottus pistilliger.	*Salmo quinnat.*
Cottus polyacanthocephalus.	*Salmo Gairdnerii.*
Cottus asper.	*Salmo paucidens.*
Aspidophorus acipenserinus.	*Salmo Scouleri.*
Hemilepidotus Tilesii.	*Salmo truppitch.*
Blepsias trilobus.	*Salmo nitidus.*
Sebastes variabilis.	*Mallotus pacificus.*
Cyprinus balteatus.	*Cyclopterus ventricosus.*
Leuciscus caurinus.	*Acipenser transmontanus.*"]
Leuciscus oregonensis.	

Histoire Naturelle des Poissons, par M. le Bon **Cuvier**, . . . ; et par M. Valenciennes, ... Tome dix-huitième. À Paris, chez P. Bertrand, . . . , 1846. [8° ed. xix, 505 pp., 2 l.; 4° ed. xviii, 375 pp., 2 l.—pl. 520–553.] Suite du livre dix-huitième.—Cyprinoïdes. Livre dix-neuvième.—Des Esoces ou Lucioïdes.

Histoire Naturelle des Poissons, par M. le Bon **Cuvier**, . . . ; et par M. Valenciennes, ... Tome dix-neuvième. À Paris, chez P. Bertrand, . . . , 1846. [8° ed. xix, 544 pp., 3 l.; 4° ed. xv, 391 pp., 2 l.—pl. 554–590.] Suite du livre dix-neuvième.—Brochets ou Lucioïdes. Livre vingtième.—De quelques familles* de Malacoptérygiens, intermédiaires entre les Brochets et les Clupes.

[No west-coast species described.]

Histoire Naturelle des Poissons, par M. le Bon **Cuvier**, . . . ; et par M. Valenciennes, ... Tome vingtième. À Paris, chez P. Bertrand, . . . ; 1846. [8° ed. xviii, 472 pp., 1 l.; 4° ed. xiv, 346 pp. 1 l.—pl. 591–606.] Livre vingt et unième.—De la famille des Clupéoïdes.

1848—Historia Fisica y Politica de Chile segun documentos adquiridos en esta república durante doce años de residencia en ella y publicada bajo los auspicios del Supremo Gobierno. Por **Claudio Gay**, ciudadano Chileno, indi-

---

* The families referred to are:—Chirocentres (with the genus *Chirocentrus*); Alepocéphales (with *Alepocephalus*); Lutodeires (with *Chanos* and *Gonorhynchus*); Mormyres (with *Mormyrus*); Hyodontes (with *Osteoglossum*, *Ischnosoma*, and *Hyodon*); Butirins (with *Albula = Butirinus*); Élopiens (with *Elops* and *Megalops*); Amies (with *Amia*); Vastres ou Amies? (*Vastres*); famille particulière, ou Amies? (*Heterotis*); Erythroïdes (with *Erythrinus*, *Macrodon*, *Lebiasina*, and *Pyrrhulina*); and Ombres (with *Umbra*).

viduo de varias sociedades científicas nacionales y etrangeras. Zoolegia. Tomo segundo. Paris, en casa del autor. Chile, en el Museo de Historia Natural de Santiago. MDCCCXLVIII. [Text, 8º; atlas, fol.]

[Peces, pp. 137-370 and index.—In this work are described several species afterward discovered along the coast of California.]

1848—Thirtieth Congress—first session. | = | Ex. Doc. No. 41. | — | Notes of a military reconnoissance, | from | Fort Leavenworth, in Missouri, | to | San Diego, in California, | including part of the | Arkansas, Del Norte, and Gila Rivers. | — | By Lieut. Col. **W. H. Emory**. | Made in 1846-7, with the advanced guard of the "Army of the West." | — | February 9, 1848.—Ordered to be printed. | February 17, 1848.—*Ordered*, That 10,000 extra copies of each of the Reports of Lieu-| tenant Emory, Captain Cooke, and Lieutenant Abert, be printed for the use of the House; | and that of said number, 250 copies be furnished for the use of Lieutenant Emory, Captain | Cooke, and Lieutenant Abert, respectively. | Washington : | Wendell and Van Benthuysen, printers. | : : : : | 1848. [8º, 614 pp., 50 lith. pl. not numbered, 14 numbered, 2 sketch-maps, and 3 maps folded.]

[This work has been so badly edited that the following analysis may prove useful, and will facilitate the understanding of the work : –]

CONTENTS.

Notes | of | a military reconnoissance, | from | Fort Leavenworth, in Missouri, to San Diego, | in California, | including | part of the Arkansas, Del Norte, and Gila Rivers. | pp. 5-126, 26 lith. pl., 2 sketch-maps.

[A species of *Gila* is noticed at p. 62, and illustrated by a poor plate opposite the text. It is said:—" We heard the fish playing in the water, and soon those who were disengaged were after them. At first it was supposed they were the mountain trout, but, being comparatively fresh from the hills of Maine, I soon saw the difference."]

Histoire Naturelle des Poissons, par M. le Bon Cuvier, . . . ; et par M. **Valenciennes**, . . . Tomo vingt et unième. À Paris, chez P. Bertrand, . . . , 1848. [8º ed. xiv, 536 pp.; 4º ed. xiii (+ iii), 391 pp.— pl. 607-633.]

—————
* Pages 289-304 misnumbered 209-224.

Suite du livre vingt et unième et des Clupéoïdes.*
Livre vingt-deuxième.—De la famille des Salmonoïdes.
[No west-coast species described.]

1849—Frank Forrester's Fish and Fishing of the United States and British Provinces of North America. Illustrated from nature by the author. By **Henry William Herbert**, author of " Field Sports," " Warwick Woodlands," etc. New York, Stringer & Townsend, 222 Broadway, 1849. 8°.

Histoire Naturelle des Poissons, par M. le Bon Cuvier, . . . ; et par M. **Valenciennes**, . . . Tome vingt-deuxième. À Paris, chez P. Bertraud, . . . , 1849. [8° ed. xx, 532, (index) 91 (+1) pp.; 4° ed. xvi, 395, (index) 81 (+ 1) pp.—pl. 634–650.]
Suite du livre vingt-deuxième.—Suite de la famille des Salmonoïdes.
[No west-coast species described.]

A Monograph of the Fresh water Cottus of North America. By **Charles Girard**. Aug. 1849. < Proc. Am. Assoc. Adv. Sci., v. 2, pp. 409–411, 1850.

On the genus Cottus Auct. By **Charles Girard**. Oct. 17, 1849. < Proc. Bost. Soc. Nat. Hist., v. 3, pp. 183–190, 1849.

1850—Some additional observations on the nomenclature and classification of the genus Cottus. By **Charles Girard**. June 19, 1850. < Proc. Bost. Soc. Nat. Hist., v. 3, pp. 302–305, 1850.

1851—On a new genus of American Cottoids. By **Charles Girard**. Feb. 5, 1851. < Proc. Bost. Soc. Nat. Hist., v. 4, pp. 18–19, 1851.

Révision du genre Cottus des auteurs. Par **Charles Girard**, de l'Association américaine pour l'avancement des sciences, membre de la Société d'histoire naturelle de Boston. [1851. 4°, 28 pp.] < N. Denkschr. allg. Schweizer. Gesell. gesammt. Naturw., B. 12, 1852.

Smithsonian Contributions to Knowledge. = Contributions to the Natural History of the Fresh Water Fishes of North America. By **Charles Girard**. I. A Monograph of the Cottoids. Accepted for publication by the Smithsonian Institution, December, 1850. [Smithsonian Contributions to Knowledge,] vol. iii, art. 3. [4°, 80 pp., 3 pl.]

Description of a new form of Lamprey from Australia, with a Synopsis of the Family. By **J. E. Gray**, Esq., F. R. S., V. P. Z. S., etc. < Proc. Zool. Soc. London, part xix, pp. 235–241, plates, Pisces, iv, v, 1851.

List of the specimens of Fish in the collection of the British Museum.—Part I.—Chondropterygii.—Printed by order of the trustees. London, 1851. [12°, x, [1], 160 pp., 2 pl.]
[The name of the compiler is not published on the title-page. In the usual introduction, Mr. Gray states:—"The characters of the genera of Sharks and Rays, with their synonyms, have principally been derived from the work of Professors Müller and Henle. The specimens which were not named by those authors when engaged in their work, or by Dr. Andrew Smith, have been determined by Mr. Edward Gerrard." The responsibility of the compilation, however, apparently devolves on JOHN EDWARD GRAY. The diagnoses of the groups, and, for the most part, the synonymy of the species, are, in fact, translated or transcribed from Müller and Henle's great work on the Plagiostomes, entitled as follows:—Systematische Beschreibung der Plagiostomen von Dr. J. MÜLLER, o. ö. Professor der Anatomie und Physiologie, und Director des anatomischen Theaters und Museums in Berlin, und Dr. J. HENLE, o. ö. Professor der Anatomie und Director des anatomischen Theaters und Museums in Zürich. Mit sechzig Steindrucktafeln. Berlin, Verlag von Veit und Comp.—1841. [Folio, xxii, 200 pp., 2 l., 60 pl., mostly colored, unnumbered.] An epoch-marking work, but with no notices of Western American species.]

---

* The Notopteros are differentiated from the Clupeoïdes as a very distinct family (une famille très-distincte).

1851—Supplement to Frank Forrester's Fish and Fishing of the United States and British Provinces of North America. By **William Henry Herbert**, author of the "Field Sports of North America," "Frank Forrester and his Friends," etc. New York, Stringer & Townsend, 222 Broadway, 1851. pp. 1–86.

1853—Descriptions of some new Fishes from the River Zuñi. By **S. F. Baird** and **Charles Girard**. June 28, 1853. < Proc. Acad. Nat. Sci., vol. 6, pp. 368–369, June, 1853.

[N. g. and sp. *Gila* (n. g. 368), *Gila robusta* (369), *Gila elegans* (369), *Gila gracilis* (369).]

Descriptions of New Species of Fishes collected by Mr. John H. Clark, on the U. S. and Mexican Boundary Survey, under Lt. Col. Jas. D. Graham. By **Spencer F. Baird** and **Charles Girard**. August 30, 1853. < Proc. Acad. Nat. Sci. Phila., v. 6, pp. 387–390, August, 1853.

[N. sp. *Catostomus latipinnis* (388), *Gila Emoryi* (388), *Gila Grahami* (389), *Cyprinodon macularius* (389), *Heterondria affinis* (390), *Heterondria occidentalis* (390).]

32d Congress, | 2d session. { Senate. { Executive | No. 59. | — | Report of an Expedition | down the | Zuñi and Colorado Rivers, | by | Captain L. Sitgreaves, | Corps Topographical Engineers. | — | Accompanied by maps, sketches, views, and illustrations. | — | Washington: | Robert Armstrong, public printer. | 1853. [8°, 190 pp., 1 l., 24 pl. of scenery (pl. 1 folded), 6 pl. of mammals, 6 pl. of birds, 2 pl. of reptiles, 3 pl. of fishes, 21 pl. of botany, 1 folded map, all at end.]

Title. p. 1.
Report of the Secretary of War, communicating, [etc.] p. 3.
[Sitgreaves's report.] pp. 4–29.
Report | on | the natural history | of the | country passed over by the exploring expedition | under the command of Brevet Captain L. Sitgreaves, | U. S. Topographical Engineers, during the year 1851. | By S. W. Woodhouse, M. D., | surgeon and naturalist to the expedition. | pp. 31–40.
Zoology. | — | Mammals and Birds, by S. W. Woodhouse, M. D. | Reptiles, by Edward Hallowell, M. D. | Fishes, by Prof. S. F. Baird and Charles Girard. | pp. 41–152.
  Mammals. By S. W. Woodhouse, M. D. pp. 43–57, 6 pl. (1–6).
  Birds. By S. W. Woodhouse, M. D. pp. 58–105, 6 pl. (1–6).
  Reptiles. By Edward Hallowell, M. D. pp. 106–147, 21 pl. (1–20+ 10 a).
  Fishes. By Spencer F. Baird and Charles Girard. pp. 148–152, 3 pl. (1–3).
Botany. | — | By Professor John Torrey. pp. 153–178, 21 pls. (1–21).
Medical Report. | — | By S. W. Woodhouse, M. D. pp. 179–185.
List of illustrations. pp. 187–190.
Table of contents. [1 l.]

Extraordinary Fishes from California, constituting a new family, described by **L. Agassiz**. < Am. Journ. Sci. and Arts, (2), v. 16, pp. 380–390, Nov. 1853; also reprinted in Edinburgh New Phil. Journ., v. 57, pp. 214–227; translated in Archiv für Naturgeschichte (Berlin), Jahrg. 20, B. 1, pp. 149–162, 1853.

[Family named "Family Holconoti or Embiotocoidæ" (p. 383). N. g. and n. sp. *Embiotoca* (n. g., 386):—1. *Embiotoca Jacksoni* (387); 2. *Embiotoca Caryi* (389).]

[This article was translated into German as follows:—]

Ueber eine neue Familie von Fischen aus Californien. Von **L. Agassiz**. Aus Silliman's Amer. Journ. vol. xvi. p. 380 übersetzt. Vom Herausgeber [F. H. Troschel]. < Archiv für Naturgeschichte, 20. Jahrg., B. 1, pp. 149–162, 1854.

[This translation was followed by the following original communication, in which the systematic relations of the family were definitely determined:—]

Ueber die systematische Stellung der Gattung Embiotcca. Bemerkung zur vorigen Abhandlung. Vom Herausgeber [Dr. **F. H. Troschel**]. < Archiv für Naturgeschichte, 20. Jahrg., B. 1, pp. 163–168, 1854.

1854—The Zoology of the Voyage of H. M. S. Herald, under the comm.and of Captain Henry Kellett, R. N., C. B., during the years 1845–51.—Published under the Authority of the Lords Commissioners of the Admiralty.—Edited by Professor Edward Forbes, F. R. S. Vertebrals, including Fossil Mammals. By Sir **John Richardson**, Knt., C. B., M. D., F. R. S.—London: Lovell Reeve, 5, Henrietta street, Covent Garden.—1854. [4°, xi, vi, [1], 171 [+1] pp., 32 pl.]

Fish. pp. 156–171, and pl. xxviii, pl. xxxiii.

[Describes *Platessa stellata*, mouth of Coppermine River (164, pl. 32, f. 1–3); *Platessa glacialis*, Bathurst's Inlet (166, pl. 32); *Salmo consuetus*, Yukon River (167, pl. 32); *Salmo dermatinus*, Yukon River (169, pl. 33, f. 3–5).]

Notice of a collection of Fishes from the southern bend of the Tennessee River, in the State of Alabama. By **L. Agassiz**. < Am. Journ. Sci. and Arts, (2), v. 17, pp. 297–308, Mar. 1854; v. 17, pp. 353–369, May, 1854.

Appendix.—Additional notes on the Holconoti. pp. 365–369, May, 1854.

[N. g. and n. sp. *Embiotoca lateralis* (366), *Rhacochilus* (n. g.) *toxotes* (367), *Amphistichys* (n. g.) *argenteus* (367), *Holconotus* (n. g., 367) *rhodoterus* (368).]

[Translated as follows:—]

Nachträgliche Bemerkungen über die Holcoucti. Von Prof. **L. Agassiz**. Aus Silliman Amer. Journ. xvii. p. 365. Uebersetzt vom Herausgeber [J. H. Troschel]. < Archiv für Naturgeschichte, 21. Jahrg., B. 1, pp. 30–34, 1855.

Description of four new species of Viviparous Fishes from Sacramento River and the Bay of San Francisco. Read before the California Academy of Natural Sciences, May 15, 1854. By **W. P. Gibbons**, M. D. June 27, 1854. < Proc. Acad. Nat. Sci. Phila., v. 7, pp. 105–106, 1854.

[N. sp. *Hysterocarpus Traskii* (105), *Hyperprosopon argenteum* (105) and var. *a. punctatum* (106), *Cymatogaster aggregatus* (106), *Cymatogaster minimus* (106).]

Description of new Species of Viviparous Marine and Fresh-water Fishes, from the Bay of San Francisco, and from the River and Lagoons of the Sacramento. By **W. P. Gibbons**, M. D. [Read before the California Academy of Natural Sciences, Jan. 9th and May 15th, 22d, and 29th, 1854.] July 25, 1854. < Proc. Acad. Nat. Sci. Phila., v. 7, pp. 122–126, July, 1854.

[N. g. and n. sp. *Holconotus* (122), *H. Agassizii* (122), *H. Gibbonsii*, "Cal. Acad. of N. S." (122), *H. fuliginosus* (123), *Cymatogaster* (n. g.), *O. Larkinsii* (123), *O. pulchellus* (123), *C. ellipticus* (124), *Hysterocarpus* (n. g.), *H. Traskii* (124), *Hyperprosopon* (n. g.), *H. argenteus* (125), *P. arcuatus* (125), *Micrometrus* (n. g.), *M. aggregatus* (125), *M. minimus* (125), *Mytilophagus* (n. g.), *M. fasciatus* (125), *Pachylabrus* (n. g.), *P. variegatus* (126).]

[Translated as follows:—]

Beschreibung neuer Fische aus der Familie Holconoti aus dem Busen von San Francisco, aus dem Sacramento-Fluss und dessen Lagunen. Von **W. P. Gibbons**. Aus den Proceedings of the Acad. of nat. sc. of Philadelphia vol. vii. 1854. p. 122. übersetzt vom Herausgeber [F. H. Troschel]. < Archiv für Naturgeschichte, 21. Jahrg., B. 1, pp. 331–341, 1855.

Descriptions of new Fishes, collected by Dr. A. L. Heermann, Naturalist attached to the survey of the Pacific Railroad Route, under Lieut. R. S. Williamson, U. S. A. By **Charles Girard**. Aug. 29, 1854. < Proc. Acad. Nat. Sci. Phila., v. 7, pp. 129–140, 1854.

[N. g. and n. sp.:—1. *Centrarchus interruptus* (129), 2. *Cott. psis gulosus* (129), 3. *Aspicottus* (n. g.) *bison* (130), 4. *Leptocottus* (n. g., 130) *armatus* (131), 5. *Scorpænichthys* (n. g.) *marmoratus* (131), 6. *Sebastes auriculatus* (131), 7. *Chirus pictus* (132), 8. *Chirus guttatus* (132), 9.

Ophiodon (n. g.) *elongatus* (133), 10. *Gasterosteus Willia·.soni* (133), 11. *Gasterosteus microcephalus* (133), 12. *Atherinopsis* (n. g.) *californiensis* (134), 13. *Gobius gracilis* (134), 14. *Embiotoca lineata* (134), 15. *Amphistichus similis* (135), 16. *Amphistichus Heermanni* (135), 17. *Gila conocephala* (135), 18. *Pogonichthys inæquilobus* (136), 19. *Pogonichthys symmetricus* (136), 20. *Lavinia* (n. g.) *exilicauda* (137), 21. *Lavinia crassicauda* (137), 22. *Lavinia conformis* (137), 23. *Leucosomus occidentalis* (137), 24. *Ulupea mirab.lis* (138), 25. *Meletta cærulea* (138), 26. *Engraulis mordax* (138), 27. *Platichthys* (n. g.) *rugosus* (139), 28. *Pleuronichthys* (n. g.) *cænosus* (139), 29. *Parophrys* (n. g., 139) *vetulus* (140), 30. *Psettichthys* (n. g.) *melanostic.us* (140).]

1854—Enumeration of the species of marine Fishes, collected at San Francisco, California, by Dr. C. B. R. Kennerly, naturalist attached to the survey of the Pacific R. R. Route, under Lieut. A. W. Whipple. By **Charles Gira:d**. Aug. 29, 1854. < Proc. Acad. Nat. Sci. Phila., v. 7, pp. 141–142, Aug. 1854.

[N. g. and n. sp :—1. *Chirus constellatus* (141), 3. *Porichthys* (n. g.) *notatus* (141), 8. *Gadus proximus* (141), 10. *Psettichthys sordidus* (142).]

. Observations upon a collection of Fishes made on the Pacific coast of the U. States, by Lieut. W. P. Trowbridge, U. S. A., for the Museum of the Smithsonian Institution. By **Charles Girard**. Aug. 29, 1854. < Proc. Acad. Nat. Sci. Phila., v. 7, pp. 142–156, 1854.

[N. g. and n. sp.:—1. *Labrax nebul'fer* (142), 2. *Labrax clathratus* (143), 3. *Heterostichus* (n. g.) *rostratus* (143), 4. *Sphyræna argentea* (144), 5. *Cottopsis parvus* (144), 8. *Scorpænichth.ys lateralis* (145), 9. *Scorpæna guttata* (145), 11. *Sebastes rosaceus* (146), 12. *Sebastes fasciatus* (146), 15. *Gasterosteus plebeius* (147), 16. *Gasterosteus inopinatus* (147), 17. *Umbrina undulata* (148), 18. *Glyphisodon rubicundus* (148), 19. *Belone exilis* (149), 20. *Blennius gentilis* (149), 21. *Gunnellus ornatus* (149), 22. *Apodichthys* (n. g.) *flavidus* (150), 23. *Apodichthys violaceus* (150), 24. *Anarrhichas felis* (150), 26. *Julis modestus* (151), 29. *Embiotoca lineata* (151), 30. *Embiotoca Cassidyi* (151), 32. *Holconotus Trowbridgii* (152), 33. *Holconotus megalops* (152), 31. *Phanerodon* (n. g.) *furcatus* (153), 36. *Pogonichthys argyreiosus* (153), 37. *Fundulus parvipinnis* (154), 42. *Engraulis delicatissimus* (154), 43. *Argentina pretiosa* (150), 44. *Pleuronectes maculosus* (155), 48. *Lepadogaster reticulatus* (155), 49. *Syngnathus breviroltris* (156), 50. *Syngnathus leptorhynchus* (156).]

† Descriptions of two species of fish, believed to be new. Sept. 4, 1854. By **Wm. O. Ayres**. < Proc. Cal. Acad. Sci., v. 1, pp. 3–4, 1854; 2d ed., pp. 3–4, 1873.

[N. sp. *Labrus pulcher, Hemitripterus marmoratus*.]

† Descriptions of two new species of Sebastes. Sept. 11, 1854. By **Wm. O. Ayres**. < Proc. Cal. Acad. Sci., v. 1, pp. 5–6, 1854; 2d ed., pp. 5–6, 1873.

[N. sp. *S. nebulosus, S. paucispinis*.]

† Descriptions of new species of fish. Sept. 18, 1854. By **Wm. O. Ayres**, M. D. < Proc. Cal. Acad. Sci., v. 1, pp. 7–8, 1854; 2d ed., pp. 7–8, 1873.

[N. sp. *Sebastes ruber, Sebastes ruber* var. *parvus, Sebastes variabilis, Centrarchus maculosus*.]

† Observations on the development of Añablcps Gronovii, a viviparous fish from Surinam. By Prof. **Jeffries Wyman**. Sept. 20, 1854. < Proc. Boston Soc. Nat. Hist., v. 5, pp. 80–81, Dec. 1854.*

* Remarks in relation to the Mode of Development of Embiotocoidæ. By **Charles Girard**. Sept. 20, 1854. < Proc. Boston Soc. Nat. Hist., v. 5, pp. 81–82, Dec. 1854.

* Two new fishes, Morrhua californica and Grystes lineatus. By **Wm. O. Ayres**. Oct. 2, 1854. < Proc. Cal. Acad. Sci., v. 1, pp. 9–10, 1854; 2d ed., pp. 8–10, 1873.

[N. sp. *Morrhua californica, Grystes lineatus*.]

*See, also, Observations on the development of Anableps Gronovii (Cuv. and Val.). By Jeffries Wyman, M. D. Read Sept. 20, 1854. < Boston Journ. Nat. Hist., v. 6, pp. 432–443, pl. 17, Nov. 1854.

1854—†Descriptions of a new species of cottoid fish, and remarks on the American Acanthocotti. By **Wm. O. Ayres**, M. D. Oct. 9, 1854. < Proc. Cal. Acad. Sci., v. 1, p. 11, 1854; 2d ed., p. 11, 1873.

[N. sp. *Clypeocottus robustus* (= *Aspicottus bison* Grd.).]

† Descriptions of two new species of fish. By **Wm. O. Ayres**, M. D. Oct. 23, 1854. < Proc. Cal. Acad. Sci., v. 1, pp. 13–14, 1854; 2d. ed., pp. 12–13, 1873.

[N. sp. *Brosmius marginatus, Syngnathus griseolineatus.*]

New species of Californian Fishes, by **William O. Ayres**, M. D. Nov. 1, 1854. < Proc. Boston Soc. Nat. Hist., v. 5, pp. 94–103, Dec. 1854, and Feb. 1855.

[N. sp. *Sebastes paucispinis* (94), *Sebastes nebulosus* (96), *Sebastes ruber* (97), *Sebastes ruber* var. *parvus* (98), *Centrarchus maculosus* (99), *Morrhua californica* (100), *Labrus pulcher* (101).]

*Descriptions of the Sturgeons [Acipenser] found in our [Californian] waters. By **Wm. O. Ayres**, M. D. Nov. 27, 1854. < Proc. Cal. Acad. Sci., v. 1, p. 15, Dec. 1854; 2d ed., pp. 14–15, 1873.

[N. sp. *A. acutirostris, A. medirostris, A. brachyrhynchus.*]

Characteristics of some Cartilaginous Fishes of the Pacific coast of North America. By **Charles Girard**. Nov. 28, 1854. < Proc. Acad. Nat. Sci. Phila., v. 7, pp. 196–197, 1854.

[N. sp.:—1. *Cestracion francisci* (196), 2. *Triakis semifasciatum* (196), 3. *Spinax (Acanthias) Suckleyi* (196), 5. *Raja binoculata* (196).]

Abstract of a Report to Lieut. Jas. M. Gilliss, U. S. N., upon the Fishes collected during the U. S. N. Astronomical Expedition to Chili. By **Charles Girard**. Nov. 28, 1854. < Proc. Acad. Nat. Sci. Phila., v. 7, pp. 197–199, 1854.

[Genus *Atherinopsis* noticed, and the *Meletta cærulea* of Aug. 29, 1854, v. 7, p. 138, redescribed as a new species, under the name *Alosa musica.*]

†Descriptions of two new species of fish. By **Wm. O. Ayres**, M. D. Dec. 4, 1854. < Proc. Cal. Acad. Sci., v. 1, pp. 17–18, 1854; 2d ed., pp. 16–17, 1873.

[N. sp. *Osmerus elongatus, Mustelus felis.*]

†Descriptions of two new species of Cyprinoids. By **Wm. O. Ayres**, M. D. Dec. 11, 1854. < Proc. Cal. Acad. Sci., v. 1, pp. 18–19, 1854; 2d ed., pp. 17–18, 1873.

[N. sp. *Catostomus occidentalis, Gila grandis.*]

*Descriptions of two new Cyprinoid fish. By **Wm. O. Ayres**, M. D. Dec. 18, 1854. < Proc. Cal. Acad. Sci., v. 1, pp. 20–21, 1854; 2d ed., pp. 19–20, 1873.

[N. sp. *Lavinia gibbosa, L. compressa.*]

*Description of a new Cyprinoid fish. By **Wm. O. Ayres**, M. D. Dec. 25, 1854. < Proc. Cal. Acad. Sci., v. 1, pp. 21–22, 1854; 2d ed., pp. 20–21, 1873.

[N. sp. *Gila microlepidota.*]

A list of the Fishes collected in California, by Mr. E. Samuels, with descriptions of the new species. By **Charles Girard**, M. D. [1854.] < Boston Journ. Nat. Hist., v. 6, pp. 533–544, pl. 24–26, 1857.

1855—Synopsis of the Ichthyological Fauna of the Pacific Slope of North America, chiefly from the collections made by the U. S. Exp. Exped. under the command of Capt. C. Wilkes, with recent additions and comparisons with eastern types. By **Louis Agassiz**. < Am. Journ. Sci. and Arts, v. 19, pp. 71–99, Jan., 1855; v. 19, pp. 215–231, March, 1855.

[N. g. and n. sp. *Catostomus occidentalis* (94), *Acrocheilus* (n. g., 96) *alutaceus* (99), *Ptychocheilus* (n. g., 227), *Ptychocheilus gracilis* (229), *Ptychocheilus major* (229), *Mylocheilus* (n. g. 229) *lateralis* 231).]

1855—* On two species of Liparis. By **Wm. O. Ayres**, M. D. Jan. 8, 1855. < Proc. Cal. Acad. Sci., v. 1, pp. 23–24, Feb. 1, 1855; 2d ed., pp. 21–23, 1873.
[N. sp. *L. pulchellus, L. mucosus.*]

† Description of a new genus (Leptogunellus) and two new species of fishes. By **Wm. O. Ayres**, M. D. Jan. 22, 1855. < Proc. Cal. Acad. Sci., v. 1, pp. 25–27, 1855; 2d ed., pp. 24–25, 1873.
[N. sp. *Leiostomus lineatus, Leptogunellus gracilis.*]

† Description of a Lamprey, from the vicinity of San Francisco. By **Wm. O. Ayres**, M. D. Feb. 5, 1855. < Proc. Cal. Acad. Sci., v. 1, p. 28, Feb. 1., 1855; 2d ed., p. 27, 1873.
[N. sp. *Petromyzon plumbeus.*]

* Remarks on the fœtal Zygæna (Hammer-headed Shark). By **Jeffries Wyman**. Feb. 21, 1855. < Proc. Boston Soc. Nat. Hist., v. 5, p. 157, March, 1855.

† Description of a new generic type among fishes. By **Wm. O. Ayres**, M. D. Feb. 26, 1855. < Proc. Cal. Acad. Sci., v. 1, pp. 31–32, 1855; 2d ed., pp. 30–31, 1873.
[N. sp. *Anarrhichthys ocellatus.*]

† Description of a new species of Catastomus. By **Wm. O. Ayres**, M. D. March 5, 1855. < Proc. Cal. Acad. Sci., v. 1, pp. 32–33, 1855; 2d ed., pp. 31–32, 1873.
[N. sp. *Catostomus labiatus.*]

* Description of a new ichthyic type. By **Wm. O. Ayres**, M. D. March 12, 1855. < Proc. Cal. Acad. Sci., v. 1, pp. 33–35, 1855; 2d ed., pp. 32–34, 1873.
[N. g. and n. sp. *Mylopharodon* (n. g) *robustus.*]

* Description of a new Trout. By **W. P. Gibbons**. March 19, 1855. < Proc. Cal. Acad. Sci., v. 1, pp. 36–37, 1855; 2d ed., pp. 35–36, 1873.
[N. sp. *Salmo iridea.*]

* On specimens of Gasterosteus plebeius, Gir., brought from San José by the Rev. Mr. Douglas. By **Wm. O. Ayres**, M. D. April 2, 1855. < Proc. Cal. Acad. Sci., v. 1, p. 40, 1855; 2d ed., p. 39, 1873.

† Description of a new Platessa, and remarks on the Flatfish of the San Francisco markets. By **Wm. O. Ayres**, M. D. April 2, 1855. < Proc. Cal. Acad. Sci., v. 1, pp. 39–40, 1855; 2d ed., pp. 39–40, 1873.
[N. sp. *Platessa bilineata.*]

† Description of a new Salmo and a new Petromyzon. By **Wm. O. Ayres**. April 16, 1855. < Proc. Cal. Acad. Sci., v. 1, pp. 43–45, 1855; 2d ed., pp. 42–44, 1873.
[N. sp. *Salmo rivularis, Petromyzon ciliatus.*]

Notice upon the Viviparous Fishes inhabiting the Pacific coast of North America, with an enumeration of the species observed. By **Charles Girard**. April 24, 1855. < Proc. Acad. Nat. Sci. Phila., v. 7, pp. 318–323, 1855.
[N. g. and n. sp.:—3. *Embiotoca Webbi* (320), 5. *Embiotoca ornata* (321), 6. *Embiotoca perspicabilis* (321), 7. *Damalichthys* (n. g.) *vacca* (321), 9. *Abeona* (n. g.) *Trowbridgii* (322), 11. *Ennichthys* (n. g., 322), *Ennichthys megalops* (323), 12. *Ennichthys Heermanni* (323).]

[Translated into German by Dr. Troschel as follows:—]
Ueber die lebendig gebärenden Fische an der Westküste von Nordamerika. Von **Charles Girard**. (Proceedings of the Academy of nat. sc. of Philadelphia April 1855.) Uebersetzt vom Herausgeber [Prof. Dr. Troschel]. < Archiv für Naturgeschichte, 21. Jahrg., B. 1, pp. 342–354 [numb. 344], 1855.

1855 —† Description of a Gasterosteus believed to be new, and on the American species of the genus. By **Wm. O. Ayres.** April 30, 1855. < Proc. Cal. Acad. Sci., v. 1, pp. 47–48, 1855; 2d ed., pp. 46–47, 1873.

[N. sp. *Gasterosteus serratus*; name *Gasterosteus dekayi* proposed for *Gasterosteus biaculeatus* DeKay.]

† Description of a new species of Apodichthys. By **William O. Ayres,** M. D. May 21, 1855. < Proc. Cal. Acad. Sci., v. 1, pp. 55–56, 1855; 2d ed., pp. 54–55, 1873.

[N. sp. *Apodichthys virescens.*]

† Description of a new generic type of Blennoids. By **William O. Ayres,** M. D June 4, 1855. < Proc. Cal. Acad. Sci., v. 1, pp. 58–59, 1855; 2d ed., pp. 58–59, 1873.

[N. sp. *Cebedichthys cristagalli.*]

† Description of a new Carangoid fish. By **William O. Ayres,** M. D. July 2, 1855. < Proc. Cal. Acad. Sci., v. 1, pp. 62–63, 1855; 2d ed., p. 64, 1873.

[N. sp. *Caranx symmetricus.*]

† Description of a new species of Whiting. By **William O. Ayres,** M. D. July 16, 1855. < Proc. Cal. Acad. Sci., v. 1, p. 64, 1855; 2d ed., pp. 65–66, 1873.

[N. sp. *Merlangus productus.*]

* Description of a fish, representing a type entirely new to our waters. By **Wm. O. Ayres,** M. D. Aug. 6, 1855. < Proc. Cal. Acad. Sci., v. 1, pp. 66–67, 1855; 2d ed., p. 69, 1873.

[N. sp. *Saurus lucioceps.*]

* Description of a new species of Cramp fish. By **William O. Ayres,** M. D. Sept. 10, 1855. < Proc. Cal. Acad. Sci., v. 1, pp. 70–71, 1855; 2d ed., pp. 74–75, 1873.

[N. sp. *Torpedo californica.*]

† On a viviparous fish from Japan. By **Louis Agassiz.** Sept. 11, 1855. < Proc. Am. Acad. Arts and Sci., v. 3, p. 204, 1855.

"A Flying Fish, *Exocœtus fasciatus* Le Sueur, from the Pacific Ocean, lat. 30° 06′ N., long. 113° 02′ W. [Gulf of California], presented by Dr. **Lanszweert.**" Sept. 24, 1855. < Proc. Cal. Acad. Sci., v. 1, pp. 71–73, 1855.

i Description of a Shark of new generic type. By **Wm. O. Ayres,** M. D. Oct. 8, 1855. < Proc. Cal. Acad. Sci., v. 1, pp. 72–73, 1855; 2d ed., pp. 76–77, 1873.

[N. sp. *Notorhynchus maculatus.*]

* Remarks concerning a collection of fishes made by Lieut. W. P. Trowbridge at or near Cape Flattery, W. T. By **Wm. O. Ayres,** M. D. Oct. 22, 1855. < Proc. Cal. Acad. Sci., v. 1, p. 74, 1855; 2d ed., p. 79, 1873.

[10 species enumerated.]

† On a supposed new genus of Cottoids. By **Wm. O. Ayres,** M. D. Dec. 24, 1855. < Proc. Cal. Acad. Sci., v. 1, pp. 75–77, 1855; 2d ed., pp. 81–82, 1873.

[N. sp. *Calycilepidotus spinosus, Scorpænichthys lateralis* Grd.= *Calycilepidotus lateralis.*

1856—Contributions to the Ichthyology of the Western Coast of the United States, from specimens in the Museum of the Smithsonian Institution. By **Charles Girard**, M. D. June 24, 1856. < Proc. Acad. Nat. Sci. Phila., v. 8, pp. 131-137, 1855.

[N. g. and n. sp. *Paralabrax* (n. g., 131), *Homalopomus* (n. g.) *Trowbridgii* (132), *Oligocottus* (n. g., 132) *maculosus* (133), *Leiocottus* (n. g.) *hirundo* (133), *Arted.us* (n. g., 134), *Artedius notospilotus* (134), *Sebastes melanops* (135), *Oplopoma* (n. g.) *pantherina* (135), *Gasterosteus intermedius* (135), *Gasterosteus pugetti* (135), *Gobius Newberryi* (136), *Embiotoca argyrosoma* (136), *Coregonus Williamsoni* (136), *Platichthys umbrosus* (136), *Pleuronichthys guttulatus* (137), *Ammodytes personatus* (137), *Rhinoptera vespertil.o* (137).]

Researches upon the Cyprinoids inhabiting the fresh water Fishes of the United States of America, west of the Mississippi Valley, from specimens in the Museum of the Smithsonian Institution. By Charles Girard, M. D. Sept. 30, 1856. < Proc. Acad. Nat. Sci. Phila., v. 8, pp. 165-213, 1856.

[N. g. and n. sp. *Mylocheilus fraterculus* (169), *Catostomus* (*Acomus*, n. s. g.) *generosus* (174), *Catostomus macrocheilus* (175), *Catostomus bernardini* (175), *Algansea* (n. g.), *Algansea bicolor* (183), *Algansea obesa* (183), *Algansea formosa* (183), *Lavinia harengus* (184), *Argyreus nubilus* (186), *Argyreus osculus* (186), *Argyreus notabilis* (186), *Agosia* (n. g.), *Agosia chrysogaster* (187), *Agosia metallica* (187), *Meda* (n. g.) *fulgida* (192), *Richardsonius* (n. g.) *lateralis* (202), *Tiaroga* (n. g.) *cobitis* (204), *Tigoma* (n. g.), *Tigoma bicolor* (206), *Tigoma purpurea* (206), *Tigoma intermedia* (206), *Tigoma obesa* (206), *Tigoma Humboldti* (206), *Tigoma lineata* (206), *Tigoma gracilis* (206), *Tigoma nigrescens* (207), *Tigoma crassa* (207), *Cheonda* (n. g.), *Cheonda Oooperi* (207), *Cheonda cœrulea* (207), *Sibo.a* (n. g.) *atraria* (208), *Ptychocheilus rapax* (209), *Ptychocheilus lucius* (209), *Ptychocheilus vorax* (209).]

Notice upon the Species of the Genus Salmo of authors, observed chiefly in Oregon and California. By Charles Girard, M. D. Oct. 28, 1856. < Proc. Acad. Nat. Sci. Phila., v. 8, pp. 217-220, 1856.

[N. sp. *Salmo spectabilis* (21°), *Fario aurora* (218), *Fario argyreus* (218), *Fario stellatus* (219), *Salar Lewisi* (219), *Salar vi·ginalis* (220).]

33d Congress, | 2d Session. } House of Representatives. { Ex. Doc. | No. 97. | = | Narrative | of | the Expedition of an American Squadron | to | the China Seas and Japan, | performed in the years 1852, 1853. and 1854, | under the command of | Commodore M. C. Perry, United States Navy, | by | order of the Government of the United States. | — | Volume II. With illustrations. | — | Washington: | A. O. P. Nicholson, printer. | 1856. [4°, 4 p. l., 414 pp.; [Treaty,] 2 p. l., 14 pp.; [Index,] iii-xi pp., 1 l.]

Notes on some figures of Japanese Fish, taken from recent specimens by the artists of the U. S. Japan Expedition. By James Carson Brevoort. (pp. 253-256, pl. iii-xii.)

[Contains notice of *Ditrema* and first notice of the recognition of the affinity between the Embiotocoids of California and the Japanese genus.]

33d Congress, 2d Session, } Senate. { Ex. Doc. No. 78. | = | Reports | of | Explorations and Surveys, | to | ascertain the most practicable and economical route for a railroad | from the | Mississippi River to the Pacific Ocean | made under the direction of the Secretary of War, | in 1853-4, | according to acts of Congress of March 3, 1853, May 31, 1854, and August 5, 1854. | — | Volume V. | — | Washington : | Beverley Tucker, Printer. | 1855.

Explorations and Surveys for a railroad route from the Mississippi River to the Pacific Ocean. | War Department. | = | Routes in California, to connect with the routes near the thirty-fifth and thirty-second | parallels, explored by Lieut. R. S. Williamson, Corps Topographical Engineers, in 1853. | — | Geological report, | by | William P. Blake, | Geologist and Mineralogist of the Expedition. | [With appendix.] | — | Washington, D. C. | 1857. =

Appendix.—Article I. Notice of the fossil fishes.—By Professor **Louis Agassiz.**—(pp. 313-316, and 1 plate ("Fossils plate 1"))

1856—3d Congress, | 2d Session. } Senate. { Ex. Doc. | No. 78. = Reports | of | Explorations and Surveys, | to | ascertain the most practicable and economical route for a railroad | from the | Mississippi River to the Pacific Ocean | made under the direction of the Secretary of War, in | 1853-4, | according to acts of Congress of March 3, 1853, May 31, 1854, and August 5, 1854. | -- | Volume IV. | — | Washington: | Beverley Tucker, Printer. | 1856.

Explorations and surveys for a railroad route from the Mississippi River to the Pacific Ocean. | War Department. | = | Route near the thirty-fifth parallel, explored by Lieut. A. W. Whipple, Topographical | Engineers, in 1853 and 1854. | — | Report on the zoology of the expedition. | — | Washington, D. C. | 1856. = [17 pp., 1 l.]

No. 1.—Field notes and explanations.—By **C. B. R. Kennerly**, M. D., Physician and Naturalist to the Expedition.—pp. 5-17.

1857—The Northwest Coast; or, Three Years' Residence in Washington Territory. By **James G. Swan**. [Figure of terr. seal.] With numerous illustrations. New York: Harper & Brothers, Publishers, Franklin Square. 1857. [12°, 435 pp. (incl. 26 figs. and pl.), frontispiece, 1 map.]

[Popular notices of fishes—especially salmon and fishing for salmon—are given in chapters 3, 7, 9, and 14.]

* Account of some observations on the development of Anableps Gronovii, as compared with that of the Embiotocas of California. By **Jeffries Wyman**. Nov. 18, 1857. < Proc. Boston Soc. Nat. Hist., v. 6, p. 294, Jan. 1858.

Notice upon new Genera and new Species of Marine and Fresh-water Fishes from Western North America. By **Charles Girard**, M. D. Nov. 24, 1857. < Proc. Acad. Nat. Sci. Phila., v. 9, pp. 200-202, Nov. 1857.

[N. g. and n. sp. *Chiropsis* (n. g., 201), *Oligocottus analis* (201), *Oligocottus globiceps* (201), *Zaniolepis* (n. g.) *latipinnis* (202), *Blepsias oculofasciatus* (202).]

33d Congress, | 2d Session. } Senate. { Ex. Doc. | No. 78. | = | Reports | of | Explorations and Surveys, | to | ascertain the most practicable and economical route for a railroad | from the | Mississippi River to the Pacific Ocean. | Made under the direction of the Secretary of War, in | 1854-5, | according to Acts of Congress of March 3, 1853, May 31, 1854, and August 5, 1854. | — | Volume VI. | — | Washington : | Beverley Tucker, Printer. | 1857.

Explorations and Surveys for a Railroad Route from the Mississippi River to the Pacific Ocean. | War Department. | = | Routes in California and Oregon explored by Lieut. R. S. Williamson, Corps of Topographical | Engineers, and Lieut. Henry L. Abbot, Corps of Topographical Engineers, in 1855. | — | Zoological Report.— | Washington, D. C. | 1857. | = No. 1. Report upon Fishes collected on the Survey.—By **Charles Girard**, M. D.—pp. 9-34, with plates xxii *a*, xxii *b*, xxv *a*, xxv *b*, xl *a*, xlvi, lxii, lxvi, lxviii, lxx, lxxiv.

Report on the fauna and medical topography of Washington Territory. By **Geo. Suckley**, M. D. May, 1857. < Trans. Am. Med. Assoc., v. 10, pp. 181-217, 1857.

[Fishes noticed at pp. 202-203.]

1858—Description of several new species of Salmonidæ from the north-west coast of America. By **George Suckley**, M. D. Read December 6, 1858. < Ann. Lyc. Nat. Hist. New York, v. 7, pp. 1-10, 1862.

[N. sp. *Salmo Gibbsii* (1), *Salmo truncatus* (3), *Salmo gibber* (6), *Salmo confluentus* (8), *Salmo canis* (9).]

Ichthyological Notices, by **Chas. Girard**, M. D. Dec. 28, 1858. < Proc. Acad. Nat. Sci. Phila., vol. 10, pp. 223-225, Dec. 1858.

[§ 1-4, n. sp. "*Fario Newberrii*, or else *Salmo Newberrii*" (225).]

1858—Denkwürdigkeiten einer Reise nach dem russischen Amerika, nach Mikronesien and durch Kamtschatka. Von **F. H. v. Kittlitz.**—Erster Band [—Zweiter Band].—Gotha. Verlag von Justus Perthes. 1858. [8°, vol. i, xvi, 383 pp., 2 pl.; vol. ii, 2 p. l., 463 pp., 2 pl.]

1859—33d Congress, | 2d Session. } Senate. { Ex. Doc. | No. 78. | = | Reports | of | Explorations and Surveys, | to | ascertain the most practicable and economical route for a railroad | from the | Mississippi River to the Pacific Ocean. | Made under the direction of the Secretary of War, in | 1853–6, | according to Acts of Congress of March 3, 1853, May 31, 1854, and August 5, 1854. | — | Volume X. | — | Washington : | Beverley Tucker, Printer. | 1859.

Explorations and Surveys for a railroad route from the Mississippi River to the Pacific Ocean. | War Department. | = | Fishes: by **Charles Girard,** M. D. | — | Washington, D. C. | 1858.* = [xiv, 400 pp., with plates vii–viii, xiii–xiv, xvii, xviii, xxii c, xxvi, xxix, xxx, xxxiv, xxxvii, xl, xli, xlviii, liii, lix, lxi, lxiv, lxv, lxxi.]

[N. g. and n. sp. *Oligocottus globiceps* (58), *Nautichthys* (n. g., 74), *Amblodon saturnus* (98), *Pelamys lineolata* (106), *Trachurus boops* (108), *Ephippus zonatus* (110), *Neoclinus* (n. g., 114), *Neoclinus Blanchardi* (114), *Xiphidion* (n. g., 119), *Xiphidion mucosum* (119), *Ophidion Taylori* (138), *Paralichthys* (n. g., 146), *Tigoma egregia* (291), *Thaleichthys* (n. g., 325), *Thaleichthys Stevensii* (325), *Engraulis nanus* (335), *Engraulis compressus* (336), *Tetraodon politus* (340), *Hippocampus ingens* (342), *Syngnathus Abboti* (346), *Syngnathus arundinaceus* (346), *Raja Cooperi* (372), *Petromyzon lividus* (379), *Petromyzon astori* (380), *Ammocœtes cibarius* (383).†
As this report brings up our knowledge of the fish fauna of the Pacific coast slope of the United States to the time of its publication, and ma·ks an epoch in the ichthyography of the region in question, the species described are hereinbelow enumerated. Of the several columns, (1) the first contains the family name, (2) the second the generic, (3) the third the specific, and (4) the right hand one, the page where the species are described :—

Order I.—ACANTHOPTERI.

Peroidæ	Ambloplites	interruptus	10
	Paralabrax	nebulifer	33
		clathratus	34
Trachinidæ	Heterostichus	rostratus	36
Sphyrænidæ	Sphyræna	argentea	39
Heterolepididæ	Chiropsis	constellatus	42
		pictus	43
		guttatus	44
		nebulosus	45
	Oplopoma	pantherina	46
	Ophiodon	elongatus	48
Cottidæ	Cottopsis	asper	51
		gulosus	53
		parvus	54
	Oligocottus	maculosus	56
		analis	57
		globiceps	58
	Leptocottus	armatus	60
	Leiocottus	hirundo	62
	Scorpænichthys	marmoratus	64
	Aspicottus	bison	66
	Hemilepidotus	spinosus	68
	Artedius	lateralis	70

* General Report upon the Zoology of the several Pacific Railroad Routes. Part IV.
† *Dionda grisea* (230), "from twenty miles west of Choctaw agency", is the only other new species described.

Order IV.—PHYSOSTOMI or MALACOPTERI—Continued.

Order VII.—GANOIDEI.

Sturionidæ	Acipenser	brachyrhynchus	355
		transmontanus	355
		acutirostris	355
		medirostris	356

Order VIII.—HOLOCEPHALI.

Chimæridæ	Chimæra	Colliei	360

Order IX.—PLAGIOSTOMI.

Suborder I.—SQUALI.

Scylliodontidæ	Triakis	semifasciatus	362
Mustelidæ	Mustelus	felis	364
Cestraciontidæ	Cestracion	francisci	365
Notidanidæ	Heptanchus	maculatus	367
Spinacidæ	Acanthias	Sucklii	368

Suborder II.—RAJÆ.

Rhinobatidæ	Rhinobatius	productus	370
Turpedinidæ	Narcine	californica	371
Raiidæ	Raja	cooperi	372
	Uraptera	binoculata	373
Myliobatidæ	Rhinoptera	vespertilio	375

Order X.—DERMOPTERI.

Suborder MARSIPOBRANCHII s. CYCLOSTOMI.

Petromyzontidæ	Petromyzon	tridentatus	377
		ciliatus	378
		lividus	379
		plumbeus	380
		astori	380
	Ammocœtes	cibarius	383

Explorations and Surveys for a Railroad route from the Mississippi River to the Pacific Ocean. | War Department. | ═ | Route near the 38:h and 39th parallels, explored by Captain J. W. Gunnison, and near the 41st | parallel, explored by Lieutenant E. G. Beckwith. | — | Zoological Report.[1] | — | Washington, D. C. | 1857. | ═ | [1] The report to which the present article belongs will be found in Vol. II of the series.

No. 4. Report on Fishes collected on the Survey.—By **Charles Girard**, M. D.—(pp. 21–27, with pl. xxiii, xlix, liv, lvi, lxxiii, lxxv.) Explorations and surveys for a railroad route from the Mississippi River to the Pacific Ocean. | War Department. | ═ | Route near the thirty-fifth parallel, explored by Lieutenant A. W. Whipple, Topographical | Engineers, in 1853 and 1854. | — | Zoological Report. | — | Washington, D. C. | 1859. | ═

No. 5. Report upon Fishes collected on the Survey.—By **C. Girard**, M. D.—pp. (47–59, with pl. iii–vi, ix, x, xxi, xxiv, xxv, xxxv, xl b, lii, lvii, lviii.)

Explorations and Surveys for a Railroad Route from the Mississippi River to the Pacific Ocean. | War Department. | ═ | Routes in California, to connect with the routes near the thirty-fifth and thirty-second | parallels, explored by Lieut. R. S. Williamson, Corps of Top. Eng., in 1853. | — | Zoological Report. | — | Washington, D. C. | 1859. ═

No. 4. Report on Fishes collected on thé Survey.—By **Charles Girard**, M. D.—(pp. 83–91, with pl. ii, xii, xxii, xxvii, xxviii, xxxi, xxxvi, xxxviii, xxxix, xlvii.)

1859—On some unusual modes of gestation in Batrachians and Fishes. By **Jeffries Wyman**. < Am. Journ. Sci. and Arts, (2), v. 27, pp. 5–13, Jan., 1859; reprinted < Can. Nat., v. 5, pp. 42–49, 1860; Zoologist, v. 18, pp. 7173–7179, 1860.

Ichthyological Notices. By **Charles Girard**, M. D. < Proc. Acad. Nat. Sci. Phila., 1859.

§ 5–27, Feb. 22, 1859, v. 10, pp. 56–58, 1859.
§ 28–40, March 29, 1859, v. 10, j p. 100–104, 1859.
§ 41–59, April 26, 1859, v. 10, pp. 113–122, 1859.
§ 60–77, May 31, 1859, v. 10, pp. 157–161, 1859.

[N. sp. *Neoclinus satiricus* (§ 5, p. 56), *Myrichthys tigrinus* (§ 6, p. 58).]

† On new fishes of the Californian coast. By **Wm O. Ayres**, M. D. Oct. 17, 1859. < Proc. Cal. Acad. Sci., v. 2, pp. 25–32, 1859.

[N. sp. *Sebastes nigrocinctus, Sebastes helvomaculatus, Sebastes elongatus, Anopl: poma* (n. g.) *merlangus, Stereolepis* (n. g.) *gigas, Squatina californica, Hippoglossus californicus, Muræna mordax, Orthagoriscus analis, Julis semicinctus.*]

Catalogue of the Fishes in the British Museum. By **Albert Günther**, ... Volume first. London: printed by order of the trustees. 1859. [August.]

At first only entitled:—Catalogue of the Acanthopterygian Fishes in the collection of the British Museum. By Dr. **Albert Günther**. Volume first. Gasterosteidæ, Berycidæ, Percidæ, Aphredoderidæ, Pristipomatidæ, Mullidæ, Sparidæ. London: printed by order of the Trustees. 1859. [General title + xxxix, 524 pp.—10s.]

1860—Salmon Fishery on the Sacramento River. By **C. A. Kirkpatrick**. < Hutchings's California Magazine, v. 4, pp. 529–534, June, 1860.

† Notes on Fishes previously described in the Proceedings, with figures of seven. By **Wm. O. Ayres**, M. D. July 2, 1860. < Proc. Cal. Acad. Sci., v. 2, pp. 52–59, 1860.

[N. g. *Halias* for *Brosmius marginatus.*]

Beiträge zur Kenntniss der Gobioiden. Von **Franz Steindachner**. (Mit 1 Tafel.) < Sitzungsb. mathem.-naturw. Classe [K. Akad. Wissensch.] vom 12. Juli 1860, xlii. Band, No. 23, Sitzung vom 18. October 1860, pp. 283–292.

* Description of new fishes. By **Wm. O. Ayres**, M. D. Aug. 6, 1860. < Proc. Cal. Acad. S i., v. 2, pp. 60–64, 1860.

[N. sp. *Trichodon lineatus, Osmerus thaleichthys,* with figures.]

Catalogue of the Fishes in the British Museum. By **Albert Günther**, ... Volume second. London: printed by order of the trustees. 1860. [Sept.]

At first only entitled:—Catalogue of the Acanthopterygian Fishes in the collection of the British Museum. By Dr. **Albert Günther**, ... Volume second. Squamipinnes, Cirrhitidæ, Triglidæ, Trachinidæ, Sciænidæ, Polynemidæ, Sphyrænidæ, Trichiuridæ, Scombridæ, Carangidæ, Xiphiidæ. London: printed by order of the Trustees. 1860. [General title + xxi, 548 pp. —8s. 6d.]

[Nov. loc. *Naucrates ductor* (374), *Echeneis remosa* (378), *Echeneis naucrates* (384). N. sp. *Cottus criniger* (522), *Aspidophoroides inermis* (524).]

Reports of Explorations and Surveys to ascertain the most practicable and economical route for a Railroad from the Mississippi River to the Pacific Ocean, made under the direction of the Secretary of War, in 1853-6, &c. Vol. X. Washington, 1859. Fishes; by **Charles Girard**, M. D. Washington, D. C., 1858. [Review, by **Theodore Gill**.] < Am. Journ. Sci. and Arts, 2d series, vol. 30, pp. 277–281, Sept. 1860.

1860—36th Congress, 1st Session. } Senate. { Ex. Doc. | = | Reports | of | Explorations
and Surveys | to | ascertain the most practicable and economical route for a
railroad | from | the | Mississippi River to the Pacific Ocean. Made under
the direction of the Secretary of War, in 1853–5, according to act of Congress
of March 3, 1853, May 31, 1854, and August 5, 1854. | —Volume XII. | Book
II. | Washington : | Thomas H. Ford, Printer. 1860.

Explorations and Surveys for a Railroad route from the Mississippi River
to the Pacific Ocean. | War Department. | = | Route near the forty-
seventh and forty-n nth parallels, explored by **I. I. Stevens,** | Governor
of Washington Territory, in 1853–'55. [pp. 9–353, 70 pl.] Zoological
report.—Washington, D. C., 1860. [viii, (1), 399 pp., 47 pl.]

No. 5.—Report upon the fishes collected on the survey.—By Dr. **G.
Suckley,** U. S. A. (pp. 307–368, with pl. i, xi, xv, xvi, xix, xx, xxxii,
xxxiii, xlii, xliii, xliv, l, li, lv, lx, lxiii, lxvii, lxix, lxxii, lxxv, viz :
Chapter I. Report upon the Salmonidæ. pp. 307–349.)
Chapter II. Report upon the Fishes exclusive of the Salmonidæ.
pp. 350–368.

[N. sp. *Salmo Masoni* (345).]

[This volume also appeared with the following title-page and modifications :—]

The Natural History of Washington Territory, with much relating to Minne-
sota, Nebraska, Kansas, Oregon and California, between the thirty-sixth and
forty-ninth parallels of Latitude, being those parts of the final Reports on
the Survey of the Northern Pacific Railroad Route, containing the Climate
and Physical Geography, with full Catalogues and Descriptions of the Plants
and Animals collected from 1853 to 1857. By **J. G. Cooper,** M. D., and Dr.
G. Suckley, U. S. A., Naturalists to the Expedition. This edition contains a
new preface, giving a sketch of the explorations, a classified table of contents, and the latest additions by the authors. With fifty-five new plates
of scenery, botany, and zoology, and an isothermal chart of the route.—New
York : Baillière Brothers, 440 Broadway. [etc.] 1859. [4°. xvii, 26 + 72 +
viii, 399 pp. (+1–4 pp. betw. 368 and 369), 61 pl., 1 map.]

† Descriptions of the Californian Atherinidæ, with figures of the species. By
Wm. O. Ayres, M. D. Oct. 1, 1860. < Proc. Cal. Acad. Sci., v. 2, pp. 73–
77, 1860.

[N. sp. *Atherinops affinis, Atherinopsis tenuis,* with figures.]

† Descriptions of two new Sciænoids, with figures. By **Wm. O. Ayres,** M. D.
Nov. 5, 1860. < Proc. Cal. Acad. Sci., v. 2, pp. 77–81, 1860.

[N. g. and sp. *Johnius nobilis, Seriphus* (n. g.) *politus.*]

† Description of new Californian fishes, with figures. By **Wm. O. Ayres,** M.
D. Dec. 3, 1860. < Proc. Cal. Acad. Sci., v. 2, pp. 82–86, April, 1862.

[N. g. and sp. *Camarina* (n. g.) *nigricans, Poronotus simillimus.*]

1861—Observations on the genus Cottus, and description of two new species
(abridged from the forthcoming report of Capt. J. H. Simpson), by **Theodore Gill.** March 20, 1861. < Proc. Boston Soc. Nat. Hist., v. 8, pp. 40–42.
April, 1861.

[N. g. and n. sp. *Potamocottus* (n. g. 40), *Potamoc. ttus punctulatus*]

Description of a new species of the genus Tigoma of Girard (abridged from
the forthcoming report of Capt. J. H. Simpson), by **Theodore Gill.** March
20, 1861. < Proc. Boston Soc. Nat. Hist., v. 8, p. 42, April, 1861.

[N. sp. *Tigoma squamata.*]

1861—Notes on the described species of Holconoti, found on the western coast of North America. By **Alexander Agassiz**. March 20, 1861. < Proc. Boston Soc. Nat. Hist., v. 8, pp. 122–134, 1861.

[The number of species is reduced to 15, which are grouped under 9 genera. N. g. *Tœniotoca* > *Embiotoca lateralis; n. sp. Hyperprosopon analis,*—neither described.]

† Communication on several new generic types of fishes, *i. e.,* Podothecus, Hoplopagrus, and Stephanolepis. By **Theodore Gill**. April 16, 1861. < Proc. Acad. Nat. Sci. Phila., [v. 13], pp. 77–78, 1861.

[N. g. and sp. *Podothecus* (n. g.).]

Revision of the genera of North American Sciæninæ. By **Theodore Gill** April 30, 1861. < Proc. Acad. Nat. Sci. Phila., [v. 13], pp. 79–89, 1861.

[N. g. *Rhinoscion* (85) for *Amblodon saturnus* Grd., *Genyonemus* (87) for *Leiostomus lineatus* Ayres.]

On the Liostominæ. By **Theodore Gill**. April 30, 1861. < Proc. Acad. Nat. Sci. Phila., [v. 13], pp. 89–93, 1861.

[Remarks on *Leiostomus lineatus* (92).]

Salmonidæ of Frazer River, British Columbia. By **C. Brew**. < Edinburgh New Philos. Journ., v. 13, p. 164, 1861.

On the Haploidonotinæ. By **Theodore Gill**. May 28, 1861. < Proc. Acad. Nat. Sci. Phila., [v. 13], pp. 100–105, 1861.

[Remarks on *Amblodon saturnus* (105).]

Notices of Certain New Species of North American Salmonidæ, chiefly in the Collection of the N. W. Boundary Commission, in charge of Archibald Campbell, Esq., Commissioner of the United States, by Dr. C. B. R. Kennerly, Naturalist to the Commission. By **George Suckley**, M. D., late Assistant Surgeon, U. S. Army. Read before the New York Lyceum of Natural History, June, 1861. < Ann. Lyc. Nat. Hist. New York, v. 7, pp. 306–313, 1862.

[N. g. and sp. *Salmo Kennerlyi* (307), *Salmo brevicauda* (308), *Salmo Warreni* (308), *Salmo Bairdii* (309), *Salmo Parkei* (309), *Oncorhynchus* (n. g., 312), *Salmo Campbelli* (313).]

Notes on some genera of fishes of the western coast of North America. By **Theodore Gill**. July 30, 1861. < Proc. Acad. Nat. Sci. Phila., [v. 13], pp. 164–168, 1861.

[N. g. *Atractoperca* (164), *Archoplites* (165), *Parephippus* (165), *Hypsypops* (165), *Sebastodes* (165), *Acantholebius* (166), *Pleurogrammus* (166), *Grammatopleurus* (166), *Megalicottus* (166). *Olinocottus* (166), *Blennicottus* (166), *Anoplagonus* (167), *Brosmophycis* (168), *Hypsagonus* (167), * *Paragonus* (167).]

On new types of Aulostomatoids, found in Washington Territory. By **Theodore Gill**. July 30, 1861. < Proc Acad. Nat. Sci. Phila., [v. 13], pp. 168–170, 1861.

[N. g. and sp. *Aulorhynchus* (n. g., 169) *flavidus* (169).]

On the genus Podothecus. By **Theodore Gill**. Sept. 24, 1861. < Proc. Acad. Nat. Sci. Phila., [v. 13], pp. 258–261, Sept. 1861.

Description of a new generic type of Blennoids. By **Theodore Gill**. Sept. 24, 1861. < Proc. Acad. Nat. Sci. Phila., [v. 13], pp. 261–263, Sept. 1861.

[N. g. and sp. *Anoplarchus* (n. g., 261) *purpurescens* (262).]

1861—Catalogue of the Fishes in the British Museum. By Albert Günthei Volume third. London : printed by order of the trustees. 1861. [Oct.]

At first only entitled:—Catalogue of the Acanthopterygian Fishes in the Collection of the British Museum. By Dr. Albert Günther. Volume third. Gobiidæ, Discoboli, Oxudercidæ, Batrachidæ, Pediculati, Blenniidæ, Acanthoclinidæ, Comephoridæ, Trachypteridæ, Lophotidæ, Teuthididæ, Acronuridæ, Hoplognathidæ, Malacanthidæ, Nandidæ, Polycentridæ, Labyrinthici, Luciocephalidæ, Atherinidæ, Mugilidæ, Ophiocephalidæ, Trichonotidæ, Cepolidæ, Gobiesocidæ, Psychrolutidæ, Centriscidæ, Fistulariidæ, Mastacembelidæ, Notacanthi. London : printed by order of the Trustees. 1861. [Published in Oct. 8°. General title + xxv, 586 + x* pp.—10s. 6d.]

[N. g. and n. sp. *Cyclopterus orbis* (158), *Liparis cyclopus* (163), *Centronotus crista-galli* (289) = *Anoplarchus crista-galli* (564), *Psychrolutes* (n. g.) *paradoxus* (516).]

* Description of a new ichthyic form from the coast of Lower California. By **Wm. O. Ayres**, M. D. Dec. 1, 1861. < Proc. Cal. Acad. Sci., vol. 2, pp. 156-158, 1862.

[N. sp. *Cynoscion parvipinnis*.]

Analytical synopsis of the order Squali and revision of the nomenclature of the genera. By **Theodore Gill**. Dec. 16, 1861. < Ann. Lyc. Nat. Hist., N. Y., v. 7, pp. 368*-370*+371-408, 1862.

Squalorum generum novorum descriptiones diagnosticæ. **Theodore Gill**, auctore. Dec. 16, 1861. < Ann. Lyc. Nat. Hist. N. Y., v. 8, pp. 409-413, 1862.

1862—Description of a new species of Hemilepidotus, and remarks on the group (Temnistiæ) of which it is a member. By **Theodore Gill**. Jan. 28, 1862. < Proc. Acad. Nat. Sci. Phila., [v. 14], pp. 13-14, 1862.

[N. sp. *Hemilepidotus Gibbsii* (13).]

On the subfamily of Argentininæ. By **Theodore Gill**. Jan. 28, 1862. < Proc. Acad. Nat. Sci. Phila., [v. 14], pp. 14-15, 1862.

[N. g. *Mesopus* (14) or *Hypomesus* (15).]

Note on the Sciænoids of California. By **Theodore Gill**. Jan. 28, 1862. < Proc. Acad. Nat. Sci. Phila., [v. 14]. pp. 16-18, 1862.

[5 species enumerated.]

‡ Notice of fresh water Fishes taken in the Bay of San Francisco. By **Wm. O. Ayres**, M. D. Feb. 3, 1862. < Proc. Cal. Acad. Sci., vol. 2, p. 163, Sept. 1862.

[8 sp. specified.]

On the limits and arrangement of the family of Scombroids. By **Theodore Gill**. March 25, 1862. <Proc. Acad. Nat. Sci. Phila., [v. 14], pp. 124-127, 1862.

Description of new species of Alepidosauroi æ. By **Theodore Gill**. March 25, 1862. < Proc. Acad. Nat. Sci. Phila., [v. 14], pp. 127-132, 1862.

[N. sp. *Alepidosaurus (Caulopus) borealis* (128), *Alepidosaurus (Caulopus) serra* (129).]

Catalogue of the fishes of Lower California in the Smithsonian Institution, collected by Mr. J. Xantus. By **Theodore Gill**. Part I. March 25, 1862. < Proc. Acad. Nat. Sci. Phila., [v. 14], pp. 140-151, 1862.

On a new genus of fishes allied to Aulorhynchus, and on the affinities of the family Aulorhynchoidæ to which it belongs. By **Theodore Gill**. April 29, 1862. < Proc. Acad. Nat. Sci. Phila., [v. 14], pp. 233-261, 1862.

1862—Catalogue of the Fishes of Lower California, in the Smithsonian Institution, collected by Mr. J. Xantus. By **Theodore Gill**. Part II. April 29, 1862. < Proc. Acad. Nat. Sci. Phila., [v. 14], pp. 242–246, 1862.

Catalogue of the Fishes of Lower California, in the Smithsonian Institution, collected by Mr. J. Xantus. By **Theodore Gill**. Part III. May 27, 1862. < Proc. Acad. Nat. Sci. Phila., [v. 14], pp. 249–262, 1862.

Notice of a collection of the Fishes of California presented to the Smithsonian Institution by Mr. Samuel Hubbard. By **Theodore Gill**. June 24, 1862. < Proc. Acad. Nat. Sci. Phila., [v. 14], pp. 274–282, 1862.

[N. g. and sp. *Hypocritichthys* (n. g., 275) *analis* (275), *Brachyistius* (n. g., 275) *frenatus* (275), *Hyperprosopon Agassizii* (276), *Oxylebius* (n. g., 277) *pictus* (278), *Apodichthys sanguineus* (279), *Apodichthys inornatus* (279), *Parophrys Hubbardii* (281), *Alausa californica* (281), *Isoplagiodon* sp. (282).]

Synopsis of the species of Lophobranchiate Fishes of Western North America. By **Theodore Gill**. June 24, 1862. < Proc. Acad. Nat. Sci. Phila., [v. 14], pp. 282–284, 1862.

[N. g. and sp. *Dermatostethus* (n. g., 283) *punctipinnis* (283), *Syngnathus dimidiatus* (283, 284).]

Catalogue of the Fishes in the British Museum. By **Albert Günther**, ... Volume fourth. London: printed by order of the trustees. 1862.

Also entitled:—Catalogue of the Acanthopterygii pharyngognathi and Anacanthini in the collection of the British Museum. . . . London: printed by order of the Trustees. 1862. [8°. General title + xxi, 534 pp.—8s. 6d.]

[N. sp. *Ditrema brevipinne* (248). *Pleuronectes Franklinii* (442), *Pleuronectes digrammus* (445), *Parophrys Ayresii* (456).]

Notes on the family of Scombroids. By **Theodore Gill**. July 29, 1862. < Proc. Acad. Nat. Sci. Phila., [v. 14], pp. 328–329, 1862.

Note on some genera of Fishes of Western North America. By **Theodore Gill**. July 29, 1862. < Proc. Acad. Nat. Sci. Phila., [v. 14], pp. 329–332, 1862.

[N. g. and sp. *Eucyclogobius* (n. g., 330), *Caularchus* (n. g., 330), *Eumicrotremus* (n. g, 330) *Hypsifario* (n. g., 330), *Lepidopsetta* (n. g., 330), *Hypsopsetta* (n. g., 330), *Orthopsetta* (n. g., 330), *Uropsetta* (n. g., 330), *Hydrolagus* (n. g., 331), *Gyropleurodus* (n. g., 331), *Holorhinus* (n. g., 331), *Entosphenus* (n. g., 331). 42 genera are stated to have been added to the Californian fauna, either as entirely new or in substitution for others erroneously identified, since the publication of Girard's work.]

On the classification of the families and genera of the Squali of California. By **Theodore Gill**. Oct. 28, 1862. < Proc. Acad. Nat. Sci. Phila., [v. 14], pp. 483–501, 1862.

[N. g. and sp. *Rhinotriacis* (n. g., 486) *Henlei* (486).]

‡ Statement in regard to Sebastes rosaceus and S. ruber. By **Wm. O. Ayres**, M. D. Nov. 3, 1862. < Proc. Cal. Acad. Sci., v. 2, p. 207, January, 1863.

* Description of Fishes believed to be new. By **Wm. O. Ayres**. M. D. Nov. 3, 1862. < Proc. Cal. Acad. Sci., v. 2, pp. 209–211, January, 1863.

[N. sp. *Sebastodes flavidus, Sebastodes ovalis*.]

* Remarks in relation to the fishes of California which are included in Cuvier's genus Sebastes. By **Wm. O. Ayres**, M. D. Nov. 3, 1862. < Proc. Cal. Acad. Sci., v. 2, pp. 211–218, January, 1863.

1862—Notices of certain new species of North American Salmonidæ, chiefly in the collection of the N. W. Boundary Commission. **By George Suckley, M. D.** See 1861, June.

1863—The Resources of California, comprising Agriculture, Mining, Geography, Climate, Commerce, etc., etc. and the past and future development of the State. **By John S. Hittel.**—San Francisco: A. Roman & Company. New York: W. J. Middleton. 1863. [12°, xvi, 464 pp.]

[Zoology, chap. vi (pp. 140-146); fishing (pp. 313-317).]

List of the Fishes sent by the Museum [of Comparative Zoology] to different Institutions, in exchange for other specimens, with Annotations. **By F. W. Putnam.** < Bull. Mus. Comp. Zool., No. 1, = v. 1, pp. 2-16, March 1, 1863.

* Remarks in relation to the genus Notorhynchus. By **Wm. O. Ayres, M. D.** March 2, 1863. < Proc. Cal. Acad. Sci., v. 3, p. 15, April, 1863.

Catalogue of the Fishes of Lower California, in the Smithsonian Institution, collected by Mr. J. Xantus. By **Theodore Gill.** Part IV. March 31, 1863. < Proc. Acad. Nat. Sci. Phila., [v. 15], pp. 80-88, 1863.

Descriptions of some new species of Pediculati, and on the classification of the group. By **Theodore Gill.** March 31, 1863. < Proc. Acad. Nat. Sci. Phila., [v. 15], pp. 88-92, 1863.

On an unnamed generic type allied to Sebastes [Sebastoplus, Gill]. By **Theodore Gill.** August 25, 1863. < Proc. Acad. Nat. Sci. Phila., [v. 15], pp. 207-209, 1863.

[Contains reference to Ayres's views on the Californian *Sebastoids.*]

* Remarks on ichthyic types new to the California Coast. By **Wm. O. Ayres,** M. D. Sept. 7, 1863. < Proc. Cal. Acad. Sci., v. 3, p. 66, Nov. 1863.

[N. sp. (undescribed) *Scomberesox* n. sp., *Alopias* n. sp.]

Synopsis of the Pomacentroids of the Western Coast of North and Central America. By **Theodore Gill.** Sept. 29, 1863. < Proc. Acad. Nat. Sci. Phila., [v. 15], pp. 213-221, 1863.

Notes on the Labroids of the Western Coast of North America. By **Theodore Gill.** Sept. 29, 1863. < Proc. Acad. Nat. Sci. Phila., [v. 15], pp. 221-224. 1863.

Synopsis of the North American Gadoid Fishes. By **Theodore Gill.** Sept. 29, 1863. < Proc. Acad. Nat. Sci. Phila., [v. 15], pp. 229-242, 1863.

Descriptions of the genera of Gadoid and Brotuloid Fishes of Western North America. By **Theodore Gill.** Sept. 29, 1863. < Proc. Acad. Nat. Sci. Phila. [v. 15], pp. 242-254, 1863.

Synopsis of the family of the Lycodoidæ. By **Theodore Gill.** Sept. 29, 1863. < Proc. Acad. Nat. Sci. Phila., [v. 15], pp. 254-262, 1863.

Descriptions of the Gobioid genera of the Western Coast of Temperate North America. By **Theodore Gill.** Sept. 29, 1863. < Proc. Acad. Nat. Sci. Phila., [v. 15], pp. 262-267, 1863.

[N. g. and sp. *Coryphopterus* (n. g., 262) *glaucofrænum* (263).]

On New Genera and Species of California Fishes.—No. I. By **J. G. Cooper,** M. D. Nov. 3, 1863. < Proc. Cal. Acad. Nat. Sci., v. 3, pp. 70-77, Nov. 1863.

[N. g. and n. sp. *Dekaya* (n. g.) *anomala, Ayresia* (n. g.) *punctipinnis, Orcynus pacificus.*]

1863—Notes on the Sebastoid Fishes occurring in the Coast of California. By **Wm. O. Ayres**, M. D., C. M. D. S. Nov. 10, 1863. < Proc. Zool. Soc. London —, pp. 390–402, 1863.

On New Genera and Species of California Fishes.—No. II. By **J. G. Cooper, M. D.** Nov. 16, 1863. < Proc. Cal. Acad. Nat. Sci., v. 3, pp. 93–97, Dec. 1863.

[N. sp. *Exocœtus californicus, Urolophus Halleri.*]

Description of the genus Stereolepis Ayres. By **Theodore Gill**. Nov. 24, 1863. < Proc. Acad. Nat. Sci. Phila., [v. 15], pp. 329–330, 1863.

Description of the genus Oxyjulis Gill. By **Theodore Gill**. Nov. 24, 1863. < Proc. Acad. Nat. Sci. Phila., [v. 15], pp. 330–331, 1863.

1864 – Catalogue of the Fishes in the British Museum. By **Albert Günther**, . . . Volume fifth. London : printed by order of the trustees. 1864.

Also entitled:—Catalogue of the Physostomi, containing the families Siluridœ, Characinidœ, Haplochitonidœ, Sternoptychidœ, Scopelidœ, Stomiatidœ, in the collection of the British Museum. . . . London : published by order of the Trustees. 1864. [8°. (Including general title) xxii, 455 pp.]

Beschreibung des Heterodontus Phillipii Bl. (Cestracion Phillipii Cuv.) mit Rücksicht auf seine fossilen Verwandten. Von **Johannes Strüver** (Göttingen). Dresden, 1864. [4°. 32 pp, 2 pl.] < Verhandl. K. Leopold-Carol. Akad. der Naturf., v. 31.

On new Genera and Species of Californian Fishes.—No. III. By **J. G. Cooper**, M. D. Jan. 4, 1864. < Proc. Cal. Acad. Nat. Sci., v. 3, pp. 108–114, 1864.

[N. g. and sp. *Myxodes* (or *Gibbonsia*, n. g.) *elegans, Gillichthys* (n. g.) *mirabilis, Pteroplatea marmorata.*]

Description of a new Labroid genus allied to Trochocopus, Gthr. By **Theodore Gill**. Mar. 29, 1864. < Proc. Acad. Nat. Sci. Phila., [v. 16], pp. 57–59, 1864.

[N. g. *Pimelometopon* (58), *Sebastomus* (59), *Sebastosomus* (59).]

Note on the nomenclature of Genera and Species of the family Echeneidoidœ. By **Theodore Gill**. Mar. 29, 1864. < Proc. Acad. Nat. Sci. Phila., [v. 16], pp. 59–61, 1864.

Critical remarks on the genera Sebastes and Sebastodes of Ayres. By **Theodore Gill**. May 31, 1864. < Proc. Acad. Nat. Sci. Phila., [v. 16], pp. 145–147 1864.

[N. sp. *Sebastosomus pinniger* (147), *Sebastosomus simulans* (147).]

Second contribution to the Seacology of California. By **Theodore Gill**. May 31, 1864. < Proc. Acad. Nat. Sci. Phila., [v. 16], pp. 147–151, 1864.

[N. sp. *Mustelus californicus* (148), *Notorhynchus borealis* (150).]

†Several points in Ichthyology and Conchology, viz: Percopsis Hammondii, n. sp., Paralepidoids and Alepidosauroids, Gymnotoids, and Campeloma vice Melantho. By **Theodore Gill**. June 7, 1864. < Proc. Acad. Nat. Sci. Phila., [v. 16], pp. 151–152, 1864.

†Ayresia punctipinnis named Chromis punctipinnis *fide* Gill. By **J. G. Cooper**, M. D. July 18, 1864. < Proc. Cal. Acad. Sci., v. 3, p. 160, 1864.

1864—Note on the Paralepidoids and Microstomatoids, and on some peculiarities of Arctic Ichthyology. By **Theodore Gill.** Sept. 27, 1864. < Proc. Acad. Nat. Sci. Phila., [v. 16], pp. 187–189, 1864.

Synopsis of the Cyclopteroids of Eastern North America. By **Theodore Gill.** Sept. 27, 1864. < Proc. Acad. Nat. Sci. Phila., [v. 16], pp. 189–194, 1864.

Synopsis of the Pleuronectoids of Californian and North-western America. By **Theodore Gill.** Sept. 27, 1864. < Proc. Acad. Nat. Sci. Phila., [v. 16], pp. 194–198, 1864.

Description of a new generic type of Pleuronectoids in the Collection of the Geological Survey of California. By **Theodore Gill.** Sept. 6, 1864. < Proc. Acad. Nat. Sci. Phila., [v. 16], pp. 198–199, 1864.

[N. g. and sp. *Metoponops* (n. g., 198) *Cooperi* (199).]

Note on the family of Stichæoids. By **Theodore Gill.** Sept. 7, 1864. < Proc. Acad. Nat. Sci. Phila., [v. 16], pp. 208–211, 1864.

1865—Note on the family of Myliobatoids, and on a New species of Ætobatis. By **Theodore Gill.** April 3, 1865. < Ann. Lyc. Nat. Hist. New York, v. 8, pp. 135–138, May, 1865.

[N. sp. *Myliobatis californicus* (137), *Ætobatis laticeps* (137).]

On the Genus Caulolatilus. By **Theodore Gill.** April 25, 1865. < Proc. Acad. Nat. Sci. Phila., [v. 17], pp. 66–68, 1865.

On the Cranial Characteristics of Gadus [Microgadus] proximus, Grd. By **Theodore Gill.** April 25, 1865. < Proc. Acad. Nat. Sci. Phila., [v. 17], p. 69, 1865.

[N. g. *Microgadus.*]

Note on several Genera of Cyprinoids. By **Theodore Gill.** April 25, 1865. < Proc. Acad. Nat. Sci. Phila., [v. 17], pp. 69–70, 1865.

Some remarks on Labrus pulcher (Ayres). By **Albert Günther,** M.A., M. D., Ph. D. May 30, 1865. < Proc. Acad. Nat. Sci. Phila., [v. 17], p. 77, 1865.

On a new Generic type of Sharks. By **Theodore Gill.** Sept. 26, 1865. < Proc. Acad. Nat. Sci. Phila., [v. 17], p. 177, 1865.

[N. g. and sp. *Mieristodus* (n. g., 177) *punctatus* (177).]

Histoire naturelle des Poissons ou Ichthyologie générale par **Aug. Duméril** Professeur-administrateur au Muséum d'Histoire Naturelle de Paris.—Ouvrage accompagné de planches.—Tome premier [.] Élasmobranches [i. e.] Plagiostomes et Holocéphales ou Chimères.—Première partie [-Seconde partie]. . . . Paris. Librairie Encyclopédique de Roret, ... 1865, [Text, 2 p. l., pp. 1–352; seconde partie, 2 p. l., pp. 353–720.] [8°; atlas larger 8°, pl. 1–14, pp. 1–8.]

Vancouver Island and British Columbia. Their History, Resources, and Prospects. By **Matthew Macfie,** F. R. G. S., five years resident in Victoria, V. I. London: Longman, Green, Longman, Roberts, & Green, 1865. [8°, xx pp. (including blank leaf and frontispiece), 1 l., 574 pp., 2 maps.]

Chapter V. General Resources of Vancouver's Island. pp. 131–171.

Fisheries. pp. 163–171.

1866—Catalogue of the Fishes in the British Museum. By **Albert Günther**, ...
Volume sixth. London: printed by order of the trustees. 1866.

Also entitled:—Catalogue of the Physostomi, containing the families Salmonidæ, Percopsidæ, Galaxidæ, Mormyridæ, Gymnarchidæ, Esocidæ, Umbridæ, Scombresocidæ, Cyprinodontidæ, in the collection of the British Museum. . . . London: printed by order of the Trustees. 1866. [8° xv, 368 pp.]

[N. sp. *Salmo lordii* (148).]

The Naturalist in Vancouver Island and British Columbia. By **John Keast Lord**, F. Z. S., Naturalist to the British North American Boundary Commission. [Vignettes.] In two volumes. Vol. I [–II]. London: Richard Bentley, New Burlington Street, publisher in ordinary to Her Majesty. 1866. [2 vols., 12°. Vol. i, xiv (incl. frontisp.), 2, 358 pp., 8 pl.; vol. ii, vii (incl. frontisp.), 2, 375 pp., 5 pl.]

Volume i.

Chapter II.—Victoria—The Salmon: its haunts and habits. pp. 36–61.
Chapter III.—Fish Harvesting. pp. 62–96.
Chapter IV.—The Round-fish, Herrings, and Viviparous Fish. pp. 97–120
Chapter V.—Sticklebacks and their Nests—The Bullhead—The Rockcod—The Chirus—Flatfish. pp. 121–141.
Chapter VI.—Halibut Fishing—Dogfish—A trip to Fort Rupert—Ransoming a Slave—A promenade with a Red skin—Bagging a Chief's head—Queen Charlotte's Islanders at Nauiamo. pp. 142–174.
Chapter VII.—Sturgeon-spearing—Man-sucker—Clams. . pp. 175–198.

Volume ii.

Appendix.
Li t of Fishes collected in the Salt and Fresh Waters of Vancouver Island and British Columbia. pp. 351–356.

[In the list are enumerated species which almost certainly were not "collected" in the waters in question.]

Hr. **W. Peters** machte eine Mittheilung über Fische (*Protopterus, Auliscops, Labrax, Labracoglossa, Nemotocentris, Serranus, Scorpis, Opisthognathus, Scombresox, Acharnes, Anguilla, Gymnomurœna, Chilorhinus, Ophichthys, Helmichthys*). < Monatsberichte der Königl. Akademie der Wissenschaften zu Berlin, 1866, pp. 509–526, 1 pl.

[N. g. and sp. *Auliscops* (n. g., 510) *spinescens* (510), *Sco.r brescx brevirostris* (521).]

1867—On the identity of the genus Alepisaurus Lowe with Plagyonus Steller. By Dr. **Albert Günther**. < Ann. and Mag. Nat. Hist., (4), v. 19, pp. 185–187.

On the nourishment of the fœtus in the Embiotoco d Fishes. By **James Blake**, M. D., F. R. C. S. Jan. 21, 1867. < Proc. Cal. Acad. Nat. Sci., v. 3, pp. 314–317, Sept. 1867.

On the organs of Copulation in the Male of the Embiotocoid Fishes. By **James Blake**, M. D., F. R. C. S. Nov. 4, 1867. < Proc. Cal. Acad. Nat. Sci., v. 3, pp. 371–372, May, 1868.

1868—Some Recent Additions to the Fauna of California. By **J. G. Cooper**, M. D. Jan. 13, 1868. < Proc. Cal. Acad. Sci., v. 4, pp. 3–13, Nov. 1868.

[The number of fishes is stated (p. 3) to be 196 in 1868, against 133 known in 1862.]

Nourishment of the Fœtus in Embiotocoid Fishes. By **James Blake**, M. D., Lond., F. R. C. S. < Journ. Anat. and Physiol., v. 2, pp. 280–282.

1868—On the anal fin appendage of Embiotocoid Fishes. By James Blake, M. D., F. R. C. S., Professor of Obstetrics in Tolard Medical College, St. Francisco, California. < Journ. Anat. and Physiol., v. 3, pp. 30–32, pl. 2, figs. 1 and 2, Nov. 1868.

The Natural Wealth of California. Comprising early history; geography, topography, and scenery; climate; agriculture and commercial products; geology, zoology, and botany; mineralogy, mines, and mining processes; manufactures; steamship lines, railroads, and commerce; immigration, population and society; educational institutions and literature; together with a detailed description of each county; its topography, scenery, cities and towns, agricultural advantages, mineral resources, and varied productions. By Titus Fey Cronise. San Francisco: H. H. Bancroft & Company. 1868. [8°, xvi, 696 pp.]

    Chapter VII. Zoology. pp. 434–501.

       Fishes. [By J. G. Cooper, M. D.] pp. 487–498.

    Chapter XIII. Miscellaneous Subjects. pp. 668–684.

       Fisheries. p. 680.

[The list of fishes was evidently prepared by Dr. J. G. Cooper, although only general acknowledgment for assistance was rendered in the preface. It was acknowledged by Dr. Cooper, as author, in the communication to the California Academy of Sciences, indicated above. Inasmuch as this was intended to be a complete enumeration of the fishes of California, the names are reproduced here.]

BONY FISHES.

Percidæ	Stereolepis	gigas	487	1
	Paralabrax	nebulifer	487	2*
	. Atractoperca	clathrata	487	3*
	Archoplites	interruptus	487	3*
Latiloidæ	Caulolatilus	anomalus	487	4
Sciænidæ	Rhinoscion	saturnus	488	5
	Leiostomus	lineatus	488	6
	Umbrina	undulata	488	7
	Atractoscion	nobile	488	8
	Seriphus	politus	488	9
Chætodonidæ	Parephippus	zonatus	488	10
	Girella	nigricans	488	11
Pomacentridæ	Glyphidodon	rubicundus	488	12
	Chromis	punctip'nnis	488	13
Embiotocoidæ	Hyaterocarpus	Traskii	489	14
	Embiotoca	Jacksoni	489	15
		argyrosoma	480	16
	Tæniotoca	lateralis	480	17
	Hypsurus	Caryi	489	18
	Damalichthys	vacca	489	19
	Phanerodon	furcatus	489	20
	Cymatogaster	aggregatus	489	21
	Rhachocheilus	toxotes	489	22
	Amphistichus	argenteus	489	23
	Holconotus	bodoterus	489	24
		pulchellus	489	25
	Hyperprosopon	argenteum	489	26
		arcuatum	489	27
		punctatum	489	28
	Hypocritichtbys	analis	489	29
	Brachyistins	frenatus	489	30
	Abeoua	minima	489	31
Labridæ	Trochocopus	pulcher	480	32
	Oxyjulis	modestus	490	33
Coryphænidæ	Poronotus	simillimus	490	34
Scombridæ	Scomber	diego	490	35

* Repeated.

1868—   BONY FISHES—Continued.

	Pelamys	lineolata	489	36
	Orcynus	pacificus	489	37
	Halatractus	dorsalis	490	38
	Trachurus	symmetricus	490	39
	Paratractus	boops	490	40
	Alepidosaurus	serra	490	41
Scomberesocidæ	Belone	exilis	490	42
Sphyrænidæ	Sphyræna	argentea	490	43
Atherinidæ	Chirostoma	californiensis	490	44
		affinis	490	45
		tenuis	490	46
Exocœtidæ	Exocœtus	californicus	490	47
Chiridæ	Chirus	constellatus	491	48
		pictus	491	49
		guttatus	491	50
	Acantholebius	nebulosus	491	51
	Oplopoma	pantherina	491	52
	Anoplopoma	merlangus	491	53
Gasterosteidæ	Gasterosteus	serratus	491	54
		pleblus	491	55
		microcephalus	491	56
		Wiliamsonii	491	57
Scorpænidæ	Scorpæna	guttata	491	58
	Sebastes	nigrocinctus	491	59
		nebulosus	491	60
		auriculatus	491	61
		ruber	491	62
		ocellatus	491	63
		elongatus ·	491	64
		paucispinis	491	65
		ovalis	491	66
		flavidus	491	67
		melanops	491	68
		rosaceus	491	69
	Trichodon	lineatus	491	70
	Blepsias	trilobus ?	491	71
Cottidæ	Cottopsis	gulosus	492	72
		parvus	492	73
	Leptocottus	armatus	492	74
	Oligocottus	maculosus	492	75
		analis	492	76
		globiceps	492	77
	Leiocottus	hirundo	492	78
	Scorpænichthys	marmoratus	492	79
	Aspicottus	bison	492	80
	Hemilepidotus	spinosus	492	81
		Gibbsii	492	82
		notospilotus	492	83
	Calycilepidotus	lateralis	492	84
Blennidæ	Anarrhichthys	ocellatus	492	85
	Xiphidion	mucosum	492	86
	Lumpenus	anguillaris	492	87
	Apodichthys	flavidus	492	88
	Cebedichthys	cristagalli	492	89
		violaceus	492	90
	Gunnellus	ornatus	492	91
Blennidæ	Blennius	gentilis	492	92
	Neoclinus	Blanchardi	492	93
	Pterognathus	satiricus	492	94
	Heterostichus	rostratus	492	95
	Gibbonsia	elegans	492	96
Batrachidæ	Porichthys	notatus	492	97
Gobidæ	Lepidogobius	gracilis	492	98

BONY FISHES—Continued.

1868—

BONY FISHES—Continued.

	Pogonichthys	inæquilobus	496	162
		symmetricus	496	163
		argyrelosus	496	164

CARTILAGINOUS FISHES.*

——	Orthagoriscus	analis	497	165
——	Gastrophysus	politus	497	166
——	Hippocampus	ingens	497	167
——	Syngnathus	californiousis	497	168
		grisoolineatus	497	169
		leptorhynchus	497	170
		dimidiatus	497	171
		arundinaceus	497	172
	Dermatostothus	punctipinnis	497	173
——	Antaceus	brachyrhynchus	497	174
		acutirostris	497	175
		medirostris	497	176
——	Hydrolagus	Colliei	497	177
——	Notorhynchus	maculatus	498	178
——	Isoplagiodon	Henlei	498	179
——	Triacis	semifasciatus	498	180
——	Gyropleurodus	Francisci	498	181
——	Acanthias	Sucklii	498	182
——	Sphyra	malleus	498	183
——	Alopias	vulpes	498	184
——	Rhina	californica	498	185
——	Rhinobatus	productus	498	186
——	Rhinoptera	vespertilio	498	187
——	Uraptera	binoculata	498	188
——	Torpedo	californica	498	189
——	Urolophus	Halleri	498	190
	Pteroplatea	marmorata	498	191
	Trygon	——?	498	192
——	Lampetra	plumbea	498	193
	Entosphenus	epihexodon	498	194
		ciliatus	498	195
——	Branchiostoma	——?	498	196

1868—Catalogue of the Fishes in the British Museum. By **Albert Günther**, . . . Volume seventh.—London : printed by order of the trustees. 1868.

Also entitled :—Catalogue of the Physostomi, containing the families Heteropygii, Cyprinidæ, Gonorhynchidæ, Hyodontidæ, Osteoglossidæ, Clupeidæ, Chirocentridæ, Alepocephalidæ, Notopteridæ, Halosauridæ, in the collection of the British Museum. . . . London : printed by order of the Trustees. 1868. [8°, xx, 512 pp.]

1870—Alaska and Its Resources. By **William H. Dall**, Director of the Scientific Corps of the late Western Union Telegraph Expedition. Boston : Lee and Shepard. 1870. [8°, xii, 628 pp, 15 pl., 1 map.]

Part II.

Chapter VI. Fisheries, Fur Trade. and other resources not previously mentioned. pp. 481–505.

Appendix.

Appendix G. Natural History. pp. 576–594.

List of the fishes of Alaska. p. 579.

Marine Fishes. p. 579.

Fresh-water fishes of the Yukon. p. 579.

[The list is very imperfect.]

---

* No families are recognized among the so-called cartilaginous fishes. These are indicated by the present writer by the lines in the family column.

1870—Mackerel-catching. [By **John C. Cremony.**] < Overland Monthly, v. 4, pp. 161-168, Feb. 1870.

The Pacific Coast Cod-fishery. [By Capt. **C. M. Scammon.**] < Overland Monthly, v. 4, pp. 436-440, May, 1870.

Catalogue of Fishes in the British Museum. By **Albert Günther,** . . . Volume eighth. London: printed by order of the trustees. 1870.
Also entitled:—Catalogue of the Physostomi, containing the families Gymnotidæ, Symbranchldæ, Murænidæ, Pegasidæ, and of the [orders] Lophobranchii, Plectognathi, [and subclasses] Dipnoi, Ganoidei, Chondropterygii, Cyclostomata, Leptocardii, in the British Museum. . . . London: printed by order of the Trustees. 1870. [8°, xxv, 549 pp.]
[Sp. new to coast:—*Galeus canis* (379). N. g. *Ichthyomyzon* (506).]

Über einige Pleuronectiden, Salmoniden, Gadoideu und Blenniiden aus der Decastris-Bay und von Viti-Levu. Von **Franz Steindachner** und weil. Prof. Dr. **Rudolph Kner.** < Sitzb. K. Akad. Wissensch., B. 61, Abth. i, pp. 421-447, pl. 1, 1870.
[7 species identified as common to Decastris Bay and the American coast]

Histoire naturelle des Poissons ou Ichthyologie générale par Aug. **Duméril** [,] Membre de l'Institut [,] professeur-administrateur au Muséum d'Histoire Naturelle de Paris.—Ouvrage accompagné de planches.—Tome second [.] Ganoïdes, Dipnés, Lophobranches. . . . . 1870.—Paris [,] Librairie Encyclopédique de Roret, ... 1870. [4 juin.—Text, 8°, 2 p. l., 624 pp.; Atlas, larger 8°, pl. 15-26, pp. 9-12, with half title.]

1871—The Food Fishes of Alaska. By **William Healy Dall.** < Rep. Comm. Agric., 1870, pp. 375-392, 1871.
[14 species specified: no new species described.]

† Remarks on the mode of attack of the Thrasher Shark. By **George Davidson.** July 11, 1870. < Proc. Cal. Acad. Sci., v. 4, p. 127, April, 1871

1872—Notice of an apparently new marine auimal from the Northern Pacific. By **P. L. Sclater,** M. A., Ph. D., F. R. S., Secretary of the Zoological Society of London. < Rep. 42d meeting Brit. Assoc. Adv. Sc., Aug. 1872, Tr. Sec., pp. 140-141.

Notice of a supposed new marine animal from Washington Territory, northwest America. [By **P. L. Sclater.**] < Nature, v. i, p. 436, Sept. 26, 1872.
[The supposed new animal was represented by "several specimens which at first sight appeared to resemble long thin peeled white willow-wand more than anything else." Mr. Sclater, in the first instance, " was inclined to regard them as possibly bones of one of the gigantic rays," and afterwards (when he had been told what they were) "as the hardened notochord of a low organized fish." They were, in truth, the axial skeletons of Pennatulid zoophytes ! ! !

Über eine neue Gattung von Fischen aus der Familie der Cataphracti Cuv., Scombrocottus salmoneus, von der Vancouvers-Insel. Von **W. C. H. Peters.** < Monatsb. K. Preuss. Akad. Wissensch. Berlin, pp. 568-570, 1872.
[N. g. and sp. *Scombrocottus* (n. g., 568) *salmoneus* (569).]

Report of the Commissioners of Fisheries of the State of California for the years 1870 and 1871. Sacramento: T. A. Springer, State printer. 1872. [8°, col. title, 24 pp.]

1872—Arrangement of the families of Fishes, or classes Pisces, Marsipobranchii, and Leptocardii. Prepared for the Smithsonian Institution. By **Theodore Gill**, M. D., Ph. D. Washington: published by the Smithsonian Institution. November, 1872. (Smithsonian Miscellaneous Collections. 247.) [8°, xlvi, 49 pp.]

42d Congress, 2d session. | Senate. | Ex. Doc. No. 34. | Message | from the | President of the United States, | communicating, | in compliance with a resolution of the 19th of January, 1869, information | in relation to the resources and extent of the fishing-grounds of the North | Pacific Ocean, opened to the United States by the treaty of Alaska. [Washington: Government Printing Office. 1872.—8°, 85 pp.]

On p. 2 entitled "The Fisheries and Fishermen of the North Pacific." By **Richard D. Cutts**.

Preliminary Report of the United States Geological Survey of Wyoming, and portions of contiguous Territories, (being a second [really fourth] annual report of progress,) conducted under authority of the Secretary of the Interior, by F. V. Hayden, United States Geologist.—Washington: Government Printing Office. 1872. [8°, 511 pp.]

Part IV. Special Reports.

VII. On the Fishes of the Tertiary Shales of Green River, Wyoming Territory. By Prof. **E. D. Cope**. pp. 425–431.

VIII. Recent Reptiles and Fishes. Report on the Reptiles and Fishes, obtained by the Naturalists of the Expedition. By **E. D. Cope**, A. M. pp. 432–442.

Preliminary Report of the United States Geological Survey of Montana, and portions of adjacent Territories; being a fifth annual report of progress. By F. V. Hayden, United States Geologist.—Conducted under authority of the Secretary of the Interior.—Washington: Government Printing Office. 1872. [8°, i–vi, 3–538 pp. (with 64 fig.), 2 pl., 5 maps folded.]

Part IV. Zoology and Botany.

VI. Report on the Recent Reptiles and Fishes of the Survey, collected by Campbell Carrington and C. M. Dawes. By **E. D. Cope**, A. M. pp. 467–476.

1873—A contribution to the Ichthyology of Alaska. By **E. D. Cope**. Jan. 17, 1873. < Proc. Am. Phil. Soc. Phila., v. 13, pp. 24–32, 1873. [Extras, March 11, 1873.]

[17 species enumerated ; n. sp. *Salmo tudes, Spratelloides bryoporus, Xiphidium cruorcum, Centronotus lætus, Chirus balias, Chirus ordinatus, Chirus trigrammus, Ammodytes alascanus, Gadus periscopus, Gadus auratus, Bathymaster signatus, Pleuronectes arcuatus.*]

Note on the Scombrocottus salmoneus of Peters, and its identity with Anoplopoma fimbria. By **Theodore Gill**, M. D. March 17, 1873. < Proc. Cal. Acad. Sci., v. 5. pp. 56–57, 1873 (April); reprinted. < Ann. and Mag. Nat. Hist., (4), v. 12, pp. 74–75, Sept. 1873.

* The first shad (Alausa præstabilis DeKay) caught in the waters of California. By **S. R. Throckmorton**. May 5, 1873. < Proc. Cal. Acad. Sci., v. 5, p. 85, May, 1873.

* On the introduction of exotic Food Fishes into the waters of California. By **S. R. Throckmorton**. May 5, 1873. < Proc. Cal. Acad. Sci., v. 5, pp. 86–88, May, 1873.

United States Commission of Fish and Fisheries.—Part I.—Report on the condition of the sea-fisheries of the south coast of New England in 1871 and 1872. By **Spencer F. Baird**, Commissioner.—With supplementary papers.—Washington: Government Printing Office. 1873. [8°, xlvii, 852 pp., 40 pl., with 38 l. explanatory (to pl. 1–38), 1 folded map.]

1873—

Notes on Liparis and Cyclopterns. By F. W. Putnam. August, 1873. < Proceedings of the American Association for the Advancement of Science, vol. 22, B, pp. 335–340, June, 1874.

1873—Annual Record of Science and Industry for 1872. Edited by **Spencer F. Baird**, with the assistance of eminent men of science.—New York: Harper & Brothers, Publishers, Franklin Square. 1873. [12°.]

    I. Pisciculture and the Fisheries.

        Fish Culture in California, pp. 407, 408.

        Report of California Fish Commissioners, p. 408, 409.

        Stocking California waters with Trout, p. 409.

        Transporting Black Bass to California, p. 409.

        Transferring Shad to the Sacramento River, p. 430.

        Stocking California with Shad, p. 430.

        Oil-works on Unalaschka, p. 436.

        Spawning of Cod-fish in Alaska, p. 436.

        Cod-fishing in the Shumagin Islands, p. 436.

        Salmon Fisheries in the Columbia River, p. 440.

        Capture of Sacramento Salmon with the Hook, p. 441.

        Fisheries of the Shumagin Islands, p. 444.

        Peculiarities of Reproduction of California Salmon, pp. 445, 446.

        Alleged Discovery of Young Shad in the Sacramento River, p. 447.

Report on the Prybilov Group or Seal Islands of Alaska. By **Henry W. Elliott**, Assistant Agent Treasury Department. Washington: Government Printing Office. 1873. [4to, 16¼ folios, not paged, with text parallel with back, and extending from bottom to top, 50 phot. pl.]

    Chapter VIII. Fish and Fisheries.

    See. also. 1875.

874— *Note on Subterranean Fishes in California. By **A. W. Chase**. <Am., Journ. Sc. and Arts (3), v. 7, p. 74, Jan., 1874; Forest and Stream, v. 2, p. 70, March 12, 1874.

† On the edible qualities of the Sacramento Salmon. By **Livingston Stone**. <Forest and Stream, v. 1, p. 331, Jan. 1, 1874.

Preparing Salmon on the Columbia River. By **Charles Nordhoff**. <Forest and Stream, v. 1, p. 397, Jan. 29, 1874. (From Harper's New Monthly Magazine.)

Salmon-fishing on the Novarro. [By **Thomas Bennett**.] <Overland Monthly, v. 12, pp. 119–124, Feb., 1874; Forest and Stream, v. 2, p. 29, Feb. 19, 1874.

Is the Yellow Perch (*Perca flavescens*) a good fish to introduce into California? [By **Livingston Stone**.] <Forest and Stream, v. 2, p. 84, March 19, 1874.

On the Plagopterinæ and the Ichthyology of Utah. By **Edward D. Cope**, A.M. Read before the American Philosophical Society, March 20, 1874. <Proc. Am. Phil. Soc. Phila., v. 14, pp. 129–139, 1874.

    [N. g. and n. sp. *Plagopterus* (n. g., 130), *argentissimus* (130), *Lepidomeda* (n. g., 131), *Lepidomeda vittata* (131), *Lepidomeda jarrovii* (132), *Clinostomus tænia* (133), *Rhinichthys henshavii* (133), *Hybopsis timpanogensis* (134), *Minomus platyrhynchus* (134), *Minomus jarrovii* (135), *Ceratichthys ventricosus* (136), *Myloleucus parovanus* (136), *Clinostomus phlegethontis* (137), *Uranidea vheeleri* (138).]

Geographical and Geological Explorations and surveys west of the 100th Meridian. First Lieutenant G. M. Wheeler, Corps of Engineers, U. S. A., in charge.

    On the Plagopterinæ and the Ichthyology of Utah. By **Edward D. Cope**, A. M.—Reprinted from the Proceedings of American Philosophical Society of Phila. Philadelphia: McCalla & Stavely, Prs., 237–9 Dock street. 1874. [8°, 14 pp.]

1874—The Introduction of Eastern Fish into the waters of the Pacific Slope, together
with an account of operations at the United States Salmon breeding Establishment on the McCloud River, California. [By **Livingston Stone.**]
< Forest and Stream, v. 2, pp. 100–102, March 26, 1874 (5½ c.).

On the Speckled Trout of Utah Lake.—Salmo virginalis, Girard. By **Dr. H. C.
Yarrow,** U. S. A. < Am. Sportsman, v. 4, pp. 68, 69, May 2, 1874.

Shad in California. [By **S. R. Throckmorton.**] < Forest and Stream, v. 3,
p. 229, May 21, 1874.

California Salmon[: its rapidity of growth. By **Livingston Stone.**]
< Forest and Stream, v. 2, p. 260, June 4, 1874.

Sports in California.—No. II.—Trout fishing at Humboldt Bay. [By Mon-
mouth.] < Forest and Stream, v. 2, pp. 273, 274 (5 c.), June 11, 1874.

Will the Columbia Salmon take the fly? [Anon.] < Am. Sportsman, v.
4, p. 165, June 13, 1874.

The Salmon Fisheries of Oregon. [By A.] < Forest and Stream, v. 2, p. 290,
June 18, 1874.

Sacramento Salmon vs. Eastern Salmon. [By **Livingston Stone.**] <Am.
Sportsman, v. 14, p. 198, June 27, 1874.

On the use of Giant Powder (Dynamite) for obtaining Specimens of Fish at
Sea. By **A. W. Chase,** U. S. Coast Survey. July 6, 1874. < Proc. Cal.
Acad. Sci., v. 5, pp. 334–337, Dec., 1874.

Ichthyic Fauna of Northwestern America. [By MORTIMER KERRY, *pseudon.*
J. M. MURPHY.] < Forest and Stream, v. 2, pp. 356, 357 (½ col.), July 16,
1874.

The Salmonidæ of the Pacific. [By MORTIMER KERRY, *pseudon.* J. M. MUR-
PHY.] < Forest and Stream, v. 2, pp. 369, 370 (6 c.), July 23, 1874.

Salmo Quinnat and Salmo Salar. [By **Charles G. Atkins.**] < Forest and
Stream, v. 2, pp. 388, 389 (2 c.), July 30, 1874.

Eastern Fish in California. What they are and what was done with them.
[From "Sacramento Record."] < Am. Sportsman, v. 4, p. 358, Sept. 5, 1874.

Oregon Salmon Fisheries. [From "Portland Oregonian."] <Am. Sportsman,
v. 4. p. 378, Sept. 12, 1874.

United States Fish Hatching in California. [Editorial.] < Forest and
Stream, v. 3, p. 81 (3 col.), Sept. 17, 1874.

Salmon Fisheries on the Columbia. <Am. Sportsman, v. 4, p. 412, Sept. 26,
1874.

The Salmon Fisheries of Oregon. < Forest and Stream, v. 3, pp. 155, 172,
Oct. 15, 22, 1874.

Annual Record of Science and Industry for 1873. Edited by **Spencer F.
Baird,** with the assistance of eminent men of science.—New York: Harper
& Brothers, Publishers, Franklin Square. 1874. [12º.]
 Shipments eastward of California Salmon, p. 433.
 Shad in the Sacramento River, p. 449.
 Shad in California waters, p. 449.
 Pacific Cod-fisheries of 1873, p. 458.
 Taking California Salmon with the Hook, p. 464.

Révision des espèces du groupe des Épinoches. Par M. **H. E. Sauvage.**
< Nouv. Arch. Mus. d'Hist. Nat., t. 10, pp. 5–32, pl. 1, 1874.

1874—Report of the Commissioners of Fisheries of the State of California for the years 1872 and 1873.—San Francisco: Francis & Valentine, printers and engravers, 517 Clay street ; 1874. [8°, 28 pp.]

United States Commission of Fish and Fisheries. Part II.—Report of the Commissioner for 1872 and 1873. A—Inquiry into the decrease of the Food-Fishes. B—The propagation of Food-Fishes in the waters of the United States. With supplementary papers. Washington: Government Printing · Office. 1874. [8°, 5 p. l., cii, (1), 808 pp., 38 pl., 3 maps folded.]

Report of the Commissioner. pp. i–xcii.

Appendix B.—The Salmon and the Trout, (species of Salmo). pp. 89–384.

III.*—On the North American species of Salmon and Trout. By **George Suckley**, Surgeon, United States Army. (Written in 1861.) pp. 91–160.

VI.—Report of operations during 1872 at the United States Salmon-Hatching Establishment on the M'Cloud River, and on the California Salmonidæ generally; with a list of specimens collected. By **Livingston Stone**. pp. 168–215.

XII.—On the Speckled Trout of Utah Lake, Salmo virginalis, Girard. By Dr. **H. C. Yarrow**, U. S. A. [etc.]. pp. 363–368.

XIII.—Miscellaneous notes and correspondence relative to Salmon and Trout. (pp. 369–379), viz:—

D—On the edible qualities of the Sacramento Salmon. [By **S. R. Throckmorton**.] pp. 373–374.

E—On the Salmon-Fisheries of the Sacramento River. By **Livingston Stone**.] pp. 374–379.

1875—Salmon-hatching on McCloud River. [By **Wm. M. Turner**.] <Overland Monthly, v. 14, pp. 79–85, Jan. 1875.

Korte Bidrag til nordisk Ichthyographie.—I. Forelobige Meddelelser om nordiske Ulkefiske. Af Dr. **Chr. Lütken**. (Meddelt den 31te Marts og 19de Maj 1875.) <Videnskabelige fra den Naturhistoriske Forening Kjobenhavn, 1876, pp. 355–388; Fr. trans., pp. 72–98, 1876.

Ichthyologische Beiträge (II). Von **Franz Steindachner**. 29. April 1875. <Sitzb. K. Akad. Wissensch., B. 71, Abth. i, pp. 443–480, 1875.

[4 Californian species mentioned.]

Ichthyologische Beiträge (III). Von **Franz Steindachner**. 17. Juni 1875. <Sitzb. K. Akad. Wissensch., B. 72, Abth. i, pp. 29–96, 1875.

[12 Californian species particularized: n. sp. *Xenichthys californiensis, Scorpis californiensis, Corvina stearnsii, Otolithus californiensis, Atherinops* n. g. or n. s. g. >*Atherinopsis affinis* Ayres.]

Description of a New Species of Trout from Mendocino County. [Typical specimen in the collection of California Academy of Natural Sciences.] By **W. R. Gibbons**, Alameda. June 22, 1875. <Proc. Cal. Acad. Sci., v. 6, pp. 142–144.

[n. sp. *Salmo mendocinensis.*]

California Fishplanting. [Signed **E. J. Hooper**.] <Forest and Stream, v. 5, pp. 19, 20, Aug. 19, 1875.

Trouting in Colorado. [Signed "Warren."] <Forest and Stream, v. 5, p. 35, Aug. 26, 1875.

Edible Fish of the Pacific. [Signed **E. J. Hooper**.] <Forest and Stream, v. 5, p. 36, Aug. 26, 1875.

Salmon Fishing east and west—How they take them in California. [Signed **Horace D. Dunn**.] <Forest and Stream, v. 5, p. 38, Aug. 26, 1875.

---

* These numbers are continuous through the volume and not subordinated to the parts.

1875—California Salmon. When to take them with a fly. [Signed "Podgers."] < Forest and Stream, v. 5, pp. 53, 54, Sept. 2, 1875.

Salmon Scores from the McCloud River. [By Sir **Rose Price.**] < Forest and Stream, v. 5, p. 54, Sept. 2, 1875.

Fishing in Montana. [Signed **A. B. Keeler.**] < Forest and Stream, v. 5, p. 54, Sept. 2, 1875.

The Speckled Beauties [*Salmo fontinalis*] in Colorado. [From "Denver News."] < Rod and Gun, v. 6, p. 348, Sept. 4, 1875.

Fishing in the McCloud River. [By Sir **Rose Price.**] < Rod and Gun, v. 6, p. 362, Sept. 11, 1875.

Carp in California. [By **E. J. Hooper.**] < Forest and Stream, v. 5, p. 115, Sept. 30, 1875.

California Angling. [By **E. J. Hooper.**] < Forest and Stream, v. 5, p. 133, Oct. 7, 1875.

Flora and Fauna of California. [By **W. M. Hinckley.**] < Forest and Stream, v. 5, p. 146, Oct. 14, 1875.

Lake Tahoe, Cal. Its Scenery and Trout Fishing. [By **E. J. Hooper.**] < Forest and Stream, v. 5, p. 151, Oct. 14, 1875.

Shipments of California Salmon eggs. [By **Livingston Stone.**] < Forest and Stream, v. 5, p. 179, Oct. 28, 1875.

Sea and Bay Fishing in California.—Wonders of the deep. [By **E. J. Hooper.**] < Forest and Stream, v. 5, pp. 197, 198, Nov. 4, 1875.

Illegal traffic in Salmon. < Forest and Stream, v. 5, p. 217, Nov. 11, 1875. [From *San Francisco Daily Evening Post*.]

Progress of Fish-culture in California. [By **E. J. Hooper.**] < Forest and Stream, v. 5, pp. 19.—227, Nov. 18, 1875.

The Oregon Salmon Fisheries. [*Anon.*] < Forest and Stream, v. 5, p. 230, Nov. 18, 1875.

Comparative size of Trout in Europe and America. [By **S. C. C.** i. e. **Clarke.**] < Forest and Stream, v. 5, p. 230, Nov. 18, 1875.

On what do Salmon Feed? [Editorial from **E. J. Hooper's** observations.] < Forest and Stream, v. 5, p. 280, Dec. 9, 1875.

Distribution of California Ova. < Forest and Stream, v. 5, p. 291, Dec. 16, 1875.

Ichthyologische Beiträge (IV). Von **Franz Steindachner.** 16. December, 1875. < Sitzb. K. Akad. Wissensch., B 72, Abth. i, pp. 551–616, 1875. [2 west-coast species described.]

Truckee River Trout. [*Anon.*] < Forest and Stream, v. 5, p. 308, Dec. 23, 1875.

What do Salmon eat? [By **R. Tallant.**] < Forest and Stream, v. 5, p. 308, Dec. 23, 1875.

Annual Record of Science and Industry for 1874. Edited by **Spencer F. Baird,** with the assistance of eminent men of science.—New York: Harper & Brothers, Publishers, Franklin Square. 1875. [12°.]
> J. Pisciculture and the Fisheries, pp. 419–428.
>> Alaska Cod-fisheries in 1873. p. 424.
>> Stocking a pond in Utah with Eels. p. 428.
>> Destruction of Fish on the Oregon coast with nitro-glycerine, p. 428.

1875—A report on the condition of affairs in the Territory of Alaska. By **Henry W.** **Elliott**, special agent of the Treasury Department.—Washington: Government Printing Office. 1875. [8°, 277 pp.]

Chapter VIII.—Fish and Fisheries. The Fisheries of Alaska. pp. 165–167.

[This is essentially a second edition of the report of Mr. Elliott, published in 1873.]

Department of the Interior.—Bulletin of the United States Geological and Geographical Survey of the Territories. F. V. Hayden, United States Geologist-in-Charge. 1874 and 1875. Vol 1.—Washington: Government Printing Office. 1875. [8°, xiii pp.+28 pp.+77 pp.+499 pp.+19 ll. unpaged, 26 pl., 3 maps, 1 woodcut.]

[Consisting of the separately paged Bulletins Nos. 1, 2, "First Series," and of the continuously paged Bulletins Nos. 1 to 6 inclusive, "Second Series," furnished with xiii pp. extra (title, table of contents, etc.). The distinction "Series" is not maintained after No. 6, which completes vol. 1.]

First Series, 1874.

No. 2. [8°, 77 pp., 1.]

Review of the Vertebrata of the Cretaceous Period, found west of the Mississippi River. By **Edward D. Cope**, A. M. pp. 5–48.

Supplementary Notices of Fishes from the Freshwater Tertiaries of the Rocky Mountains. [By **Edward D. Cope**, A. M.] pp. 49–51.

Second Series, 1875–1876.

No. 1. [8°, 47 pp.]

On the Fishes of the Tertiary Shales of the South Park [Colorado]. By **E. D. Cope**, A. M. pp. 3–5.

La Chasse aux animaux marins et les pêcheries chez les Indigènes de la côte nord-ouest d'Amérique, par **m. Alph. Pinart**.—Boulogne-sur-mer, Imp. de Charles Aigre, 4, Rue des Vieillards. 1875. [8°, 15 pp.]

Engineer Department, United States Army.—Report upon Geographical and Geological Explorations and Surveys west of the One Hundredth Meridian, in charge of First Lieut. G. M. Wheeler, Corps of Engineers, U. S. Army, under the direction of Brig. Gen. A. A. Humphreys, Chief of Engineers, U. S. Army. Published by authority of Hon. Wm. W. Belknap, Secretary of War, in accordance with acts of Congress of June 23, 1874, and February 15, 1875. In six volumes, accompanied by one topographical and one geological atlas.—Vol. V.—Zoology.—Washington: Government Printing Office. 1875. [4°.]

Chapter VI.—Report | upon | the collections of Fishes | made in portions of | Nevada, Utah, California, Colorado, New Mexico, and Arizona, | during | the years 1871, 1872, 1873, and 1874. | By | Prof. **E. D. Cope** and Dr. **H. C. Yarrow**.=pp. 635–703, pl. 26–32.

Appendix.—Description of a Mugiloid Fish from the Mesozoic Strata of Colorado [Syllæmus latifrons, Cope], pp. 701–703.

[N. sp. *Apocope couesii*, Yarrow (p. 648, pl. 27, f. 2), *Gila nigra*, Cope (p. 663, pl. 30, f. 3), *Gila seminuda*, Cope and Yarrow (p. 666, pl. 31, f. 2); *Hyborhynchus siderius*, Cope (p. 670, pl. 31, f. 6,) *Gila ardesiaca* (p. 660, pl. 30, f. 1), *Gila seminuda* (p. 666, pl. 31, f. 1), *Pantosteus*, Cope (n. g., p. 673), *Catostomus fecundus* (p. 678, pl. 32, f. 1).

"The most extended list is that of the Colorado basin" (p. 699):—

Cyprinidæ	Plagopterus	argentissimus	640
	Meda	fulgida	642
	Lepidomeda	vittata	642
		jarrovii	643
	Ceratichthys.	squamilentus	000
		oscula	647
	Apocope	couesii	648
		ventricosa	648

1875—

	Gila	egregia	662
		nigra	663
		robusta	663
		elegans	664
		gracilis	665
		grahamii	665
		nacrea	666
		seminuda	666
		emorii	667
	Hyborhynchus	siderius	670
Catostomidæ	Pantosteus	bardus	673
		delphinus	673
	Catostomus	insigne	676
		discobolus	677
	Ptychostomus	congestus	680
Coregonidæ	Coregonus	villiamsonii	682
Salmonidæ	Salmo	pleuriticus	693
Cyprinodontidæ	Girardinus	sonoriensis	695
Cottidæ	Uranidea	vheelerii	696

" The following species are those of the basin of Utah, whether from tributaries of the Great Salt Lake or not " (p. 700) :—

	Cyprinidæ	Apocope	carringtonii	645
			henshavii	645
			vulnerata	646
		Ceratichthys	biguttatus	651
		Hybopsis	timpanogensis	654
			bivittatus	000
		Gila	phlegethontis	657
			montana	657
			hydrophlox	658
			tænia	658
			egregia	662
		Siboma	atraria	667
		Myloleucus	pulverulentus	669
			parovanus	069
	Catostomidæ	Pantosteus	platyrhynchus	673
			jarrovii	074
		Catostomus	fecundus	678
	Coregonidæ	Coregonus	villiamsonii	682
	Salmonidæ	Salmo	virginalis	685
			pleuriticus	693
	Cottidæ	Uranidea	vheelerii	696
			punctulata	697

[In both of the preceding lists the enumeration is in the order of the descriptions, and *not* of the lists, which deviate considerably from the former.]

1876.—Salmon Fishing on the Mayo River, California. [Anon.] < Forest and Stream, v. 5, p. 267, 1876.

California Salmon for New Hampshire. < Forest and Stream, v. 5, p. 339, Jan. 6, 1876.

The McCloud River Reservation. [Editorial.] < Forest and Stream, v. 5, p. 355, Jan. 13, 1876.

Habits of Pacific Salmon. [By **Livingston Stone**.] < Forest and Stream, v. 5, p. 372, Jan. 20, 1876.

California Shad. [Anon.] < Forest and Stream, v. 5, p. 372, Jan. 20, 1876. (6 lines.)

Angling for Eastern Salmon (*Salmo salar*) in California waters. [*Anon.*] < Forest and Stream, v. 5, p. 390, Jan. 27, 1876.

1876—The Fisheries and Sea Lions of California. [Anon.] < Forest and Stream, v. 6, p. 387, Feb. 24, 1876.

The Natural and Economic History of the Salmonidæ—geographical distribution and artificial culture. By **Philo-Ichthyos.** < Forest and Stream, pp. 68–69 (No. 3), 106 (No. 4), 116 (No. 5), 131 (No. 6), 147 (No. 7), 164 (No. 8), 179 (No. 9).

Check List of the Fishes of the Fresh Waters of North America. By **David S. Jordan,** M. S., M. D., and **Herbert E. Copeland,** M. S. March 3, 1876. < Bulletin of the Buffalo Society of Natural Sciences, v. 2, pp. 133–164, 1876.

Viviparous Perch : [their abundance at Santa Barbara. By **H. C. Yarrow.**] < Forest and Stream, v. 6, p. 132, April 6, 1876.

Angling for Smelts in California. [By **E. J. Hooper.**] < Forest and Stream, v. 6, p. 166, April 20, 1876.

A Viviparous Perch. [Editorial.] < Forest and Stream, v. 6, p. 180, with fig., April 27, 1876.

Noget om Slægten Soulv (*Anarrhichas*) og dens nordiske Arter. Af Proffessor **Japetus Steenstrup.** Med en Tavle. < Videnskabelige Meddelelser fra den Naturhistorisk Forening i Kjobenhavn, 1876, pp. 159–202, tav. 3.

Salmon Fisheries on the Columbia River. [*Anon.* By **Barnet Phillips.**— From Appleton's Journal.] < Rod and Gun, v. 8, pp. 131–132 (5 col.), May 27, 1876, with 2 figs.

Remarks on the Various Fishes [of the family of Scorpænidæ] known as Rock Cod. By **W. N. Lockington.** July 17, 1876. < Proc. Cal. Acad. Sci., v. 7, pp. 79–82.

[N. sp. *Sebastes Ayresii* proposed as a substitute for *S. rosaceus* of Ayres, but not of Girard.]

Notes on Some California Marine Fishes, with description of a new species. By **W. N. Lockington.** July 17, 1876. < Proc. Cal. Acad. Sci., v. 7, pp. 83–88.

[N. sp. *Argyreiosus Pacificus*, Magdalena Bay.]

Ichthyologische Beiträge (V.) Von **Franz Steindachner.** 20. Juli 1876. < Sitzb. K. Akad. Wissensch., B. 74, Abth. i, pp. —, 1876.

[13 west-coast species elucidated: n. sp. *Artedius pugetensis, Siphagonus barbatus, Hypsagonus Swanii, Blakea* n. g. < *Myxodes elegans* Cooper.]

Lake Fishing in California. [By **E. J. Hooper.**] < Forest and Stream, v. 7, p. 5, Aug. 10, 1876.

Fishing this Season [summer of 1876] in California. [By **E. J. Hooper.**] < Forest and Stream, v. 7, p. 21, Aug. 17, 1876.

• Notes on Californian Fishes. By **W. N. Lockington.** September 4, 1876. < Proc. Cal. Acad. Sci., v. 7, pp. 108–110.

[N. sp. *Centropomus viridis* (provisionally named on p. 100) from Asuncion Island, Lower California.]

Connecticut River Shad for California. [By **S. F. Baird.**] < Forest and Stream, v. 7, pp. 66–67, Sept. 7, 1876.

California Shad. [Anon.] < Forest and Stream, v. 7, p. 83, Sept. 14, 1876.

The Big Fish [Salmon weighing 100 pounds] of Alaska. [Anon.] < Forest and Stream, v. 7, pp. 213–214, Nov. 9, 1876.

1876—Annual Record of Science and Industry for 1875. Edited by **Spencer F. Baird**, with the assistance of eminent men of science. New York: Harper & Brothers, Publishers, Franklin Square. 1876. [12°.]

J. Pisciculture and the Fisheries. pp. 405–440.
Salmon in the San Joaquin. pp. 430–431.
Salmon Trade of the Columbia River. pp. 431–432.
Salmon in the Sacramento River. p. 432.
United States Salmon-hatching Establishment, pp. 434–435.

Engineer Department, U. S. Army.=Report of explorations across the Great Basin of the Territory of Utah for a direct wagon-route from Camp Floyd to Genoa, in Carson Valley, in 1859. By Captain **J. H. Simpson**, Corps of Topographical Engineers, U. S. Army [now colonel of engineers, bvt. brig. gen., U. S. A.]. Made by authority of the Secretary of War, and under instructions from Bvt. Brig. Gen. A. S. Johnston, U. S. Army, commanding the Department of Utah. Washington: Government Printing Office. 1876.

Explorations across the Great Basin of Utah.=Appendix L.—Report on ichthyology. By Prof. **Theo. Gill**. pp. 383–431, 8 pl., with 8 l. explanatory.
[This chapter was written in 1861, and not subsequently revised.]

United States Commission of Fish and Fisheries. Part III.—Report of the Commissioner for 1873-4 and 1874-5. A—Inquiry into the decrease of the Food-Fishes. B—The propagation of Food-Fishes in the waters of the United States. Washington: Government Printing Office. 1876. [8°, li, 777 pp.]
Report of the Commissioner. pp. vii–xlvi.
Appendix A.—Sea fisheries and the fishes and invertebrates used as food. pp. 1–319.

V.—Account of the fisheries and seal-hunting in the White Sea, the Arctic Ocean, and the Caspian Sea. By **Alexander Schultz**. pp. 35–96.

Appendix B.—The river fisheries. pp. 321–540.

XX.—Report of operations in California in 1873. By **Livingston Stone**. pp. 377–429.

A—Clear Lake. pp. 377–381.
B—Sacramento River. pp. 382–385.
C—California aquarium-car. pp. 385–390.
D—Overland journey with live shad. pp. 390–402.
E—The McCloud River station. pp. 402–423.
F—Catalogue of collections sent to the Smithsonian Institution in 1873. pp. 424–427.
G—A list of McCloud Indian words supplementary to a list contained in the report of 1872. pp. 428–429.

XXI.—Hatching and distribution of California salmon.

A—Report on California salmon-spawn hatched and distributed. By **J. H. Slack**, M. D. pp. 431–434.
B—Hatching and distribution of California salmon in tributaries of Great Salt Lake. By **A. P. Rockwood**, Superintendent of Fisheries in Utah Territory. pp. 434–435.

XXII.—Report of operations during 1874 at the United States salmon-hatching establishment on the McCloud River, California. By **Livingston Stone**. pp. 437–478.

XXIII.—Correspondence relating to the San Joaquin River and its fishes. pp. 479–483.

1877—The Trout of Washington Territory. < Forest and Stream, v. 7, p. 413, Feb. 1, 1877.

Canned Salmon. [*Anon.*] < Forest and Stream, v. 8, p. 32, Feb. 22, 1877.

On the Genera of North American Fresh-water Fishes. [By **David S. Jordan** and **Charles H. Gilbert.** Feb. 27, 1877. .< Proc. Acad. Nat. Sc. Phila., v. —, pp. 83–104, April 17, 1877.

The Oregon Fisheries. [*Anon.* From "Pacific Life."] < Forest and Stream, v. 8, p. 49, March 1, 1877.

Fish Culture in California. < Forest and Stream, v. 8, pp. 16, 81, 207, 224. 1877.

Annual Record of Science and Industry for 1876. Edited by **Spencer F. Baird,** with the assistance of eminent men of science.—New York: Harper & Brothers, Publishers, Franklin Square. 1877. [12°.]
> I. Pisciculture and the Fisheries, pp. 385–410.
> > Biennial Report of the California Fish Commission [abstract]. pp. 401–403.
> > Cultivation of Carp in California. p. 403.

Department of the Interior: U. S. National Museum.—Bulletin of the United States National Museum.—No. 7.—Published under the direction of the Smithsonian Institution. Washington: Government Printing Office. 1877. [8°.]
> No. 7.—Contributions to the Natural History of the Hawaiian and Fanning Islands and Lower California. By **Thos. H. Streets,** M. D.

Trout Fishing in Southwestern Colorado. < Forest and Stream, v. 8, pp. 189, 190, May 3, 1877.

California Salmon Spawn for Shipment. < Forest and Stream, v. 8, p. 191, May 3, 1877.

Fishing in Lakes San Andreas and Pilercitas, California. [By **E. J. Hooper.**] < Forest and Stream, v. 8, p. 270, May 31, 1877.

Contributions to North American Ichthyology. Based Primarily on the Collections of the United States National Museum.
> A. Notes on the Cottidæ, Etheostomatidæ, Percidæ, Centrarchidæ, Aphododeridæ, Dorysomatidæ, and Cyprinidæ. With Revisions of the Genera and Descriptions of New or Little-known Species.—B. Synopsis of the Siluridæ of the Fresh Waters of North America. By **David S. Jordan.** Washington: Government Printing Office. 1877. [8°, 2 title-pages, 120 pp., 45 plates.]
> > (Bulletin of the U. S. National Museum, No. 10.)

M'Cloud and Sacramento River Trout. [From "San Francisco Pacific Life."] < Forest and Stream, v. 8, p. 299, June 14, 1877.

Stocking the Barren Waters of the Great Divide. [By J. W. B.] < Forest and Stream, v. 8, p. 400, July 19, 1877.

California Salmon in Lake Ontario. [By **Sam. Wilmot.**] < Forest and Stream, v. 8, p. 419, July 26, 1877.

†California Salmon in the James River, Va. < Forest and Stream, v. 8, p. 400, July 19, 1877.

Hatching on the Columbia. < Forest and Stream, v. 8, p. 420, July 26, 1877.

1877—The Long-Jawed Goby. By **W. N. Lockington.** <The American Naturalist, v. 11, pp. 474-478, Aug., 1877.
[An interesting account of some peculiarities in the habits of *Gillichthys mirabilis.*]

The Coregoni—Their natural history, native waters, economic value, and implements connected with their production. [*Anon.*] <Forest and Stream, v. 8, pp. 439, 440. 1877.

The Coregoni. No. Part 2. <Forest and Stream, v. 9, pp. 3, 4, Aug. 3, 1877.

A Contribution to the knowledge of Ichthyological Fauna of the Green River Shales. By **E. D. Cope.** <Bull. U. S. Geol. and Geog. Surv. Terrs., v. 3, pp. 807–819, Aug. 15, 1877.

California Salmon. [By **Emery D. Potter.**] <Forest and Stream, v. 9, p. 63, Aug. 30, 1879.

Notice of the Utah Trout in Provo rising to the fly. By W. V. S. <Forest and Stream, v. 9, p. 88, Sept. 6, 1877.

Canning Salmon. <Forest and Stream, v. 9, p. 88, Sept. 6, 1877.

Operations of the McCloud River (Cal.) Fish Hatching Establishment. <Forest and Stream, v. 9, p. 206, Oct. 13, 1877.

The Salmon Fisheries of California. <Forest and Stream, v. 9, p. 233, Oct. 25, 1877.

Salmon Trout on the Pacific Coast. <Forest and Stream, v. 9, p. 247, Nov. 1, 1877.

More about McLeod River Trout. <Forest and Stream, v. 9, p. 247, Nov. 1, 1877.

The Sportsman's Gazetteer and General Guide. The Game Animals, Birds and Fishes of North America: their habits and various methods of capture. Copious Instructions in Shooting, Fishing, Taxidermy, Woodcraft, etc. Together with A Directory to the Principal Game Resorts of the Country; illustrated with maps. By **Charles Hallock,** Editor of "Forest and Stream"; Author of the "Fishing Tourist"; "Camp Life in Florida," etc. New York: "Forest and Stream" Publishing Company, American News Company, agents. 1877. [12°, 668 pp., + 208 pp., 3 maps, 1 portrait.
Part I.—Game Animals of North America. Fishes of the Northwest, pp. 339-353. Pacific Coast Fishes, pp. 354-369.

1878—Beneficial Results of Salmon Hatching on the Sacramento River. [Editorial.] <Forest and Stream, v. 10, p. 18, Feb. 14, 1878.

Trout Fishing at Lake Bigler, California. [*Anon.*] <Forest and Stream, v. 10, p. 28, Feb. 14, 1878.

California Salmon Fishing and the Game Laws. [Signed **E. J. Hooper.**] <Forest and Stream, v. 10, p. 47, Feb. 21, 1878.

[Price of first four Shad of the season in San Francisco=$10 each.] <Forest and Stream, v. 10, p. 67, Feb. 28, 1878.

Birds and Salmon in California. [*Anon.*] <Forest and Stream, v. 10, p. 95, March 14, 1878.

Spawning of California Salmon. [Signed **B. B. Redding.**] <Forest and Stream, v. 10, p. 155, April 4, 1878.

Red Trout, or Redfish of Oregon and Idaho. [By **Charles Bendire,** U. S. A.] <Forest and Stream, v. 10, p. 156, April 4, 1878.

Carp in San Francisco. [From "Pacific Life."] <Forest and Stream, v. 10, p. 174, April 11, 1878.

1878—The Norway Trout of the Yellowstone. [*Anon.*] <Forest and Stream, v. 10, p. 175 [195], April 11, 1878.

Prof. Jordan on Characteristics of Trout. [Signed **D. S. Jordan.**] <Forest, and Stream, v. 10, p. 196, April 11, 1878.
[Contains suggestion that the original Redfish is *Hypsifario kennerlyi*.]

Manual of the Vertebrates of the Northern United States, including the District east of the Mississippi River and north of North Carolina and Tennessee, exclusive of marine species. By David **Starr Jordan**, Ph. D., M. D., Professor of Natural History in Butler University. Second Edition, revised and enlarged.—Chicago: Jansen, McClurg & Company, 1878. [12°. 407 pp., pub. May 16.]
[Contains synopsis of the American *Salmoninæ* and *Coregoninæ*.]

California Fishing Prospects. [Signed **E. J. Hooper.**] <Forest and Stream, v. 10, p. 239, May 2, 1878.

Notes on a Collection of Fishes from the Rio Grande, at Brownsville, Texas. By David **S. Jordan**, M. D. <Bull. U. S. Geol. and Geog. Surv. Terr. v. 4, [pp. 397–406, May 3;] v. 4, pp. 663–667, July 29, 1879.
[Specimens of *Hysterocarpus Traskii* indicated as an unknown Labroid form at p. 399, and described as the type of a new genus and sp. at p. 667. The specimens had been probably misplaced.]

A Catalogue of the Fishes of the Fresh Waters of North America. By **David S. Jordan**, M. D. <Bull. U. S. Geol. and Geog. Surv. Terr., v. 4, pp. 407–442, May 3, 1878.
[A simple nominal list of the fresh-water species north of the Mexican region.]

Spawning of California Brook Trout in New York. [By **James Annin, jr.**, Caledonia, N. Y.]. <Chicago Field, v. 9, p. 182, May 4, 1878. [F. M.]

California Salmon on Long Island, success of. By a member of the South Side Club. <Chicago Field, v. 9, p. 182, May 4, 1878. [F. M.]

Trout Hybrids. [Possibility of intercrossing Eastern and Californian Trouts. Editorial.] <Forest and Stream, v. 10, p. 255, May 9, 1878·

California. [Notice of distribution of land-locked Salmon and Eastern Trout by Fish Commissioners.] <Forest and Stream, v. 10, p. 255, May 9, 1878.

The heaviest American Salmon. [Notice of one weighing 82 pounds caught at the mouth of the Columbia River. By **John Goudy.**] <Forest and Stream, v. 10, p. 265, May 9, 1878.

Salmon canning on Frazer River. [By **Fred. Mather.**] <Chicago Field, v. 9, p. 196, May 15, 1878. [F. M.]

☲.—A. On the Distribution of the Fishes of the Allegheny Region of South Carolina, Georgia, and Tennessee. With Descriptions of New or Little-known Species. By David S. Jordan and Alembert W. Brayton.—B. Synopsis of the Family Catostomidæ. By David S. Jordan. Washington : Government Printing Office. 1878. (8vo, 237.)

Run of Salmon in California. Note by **A. R.** <Chicago Field, v. 9, p. 229, May 25, 1878. [F. M.]

Shad in California. Announcement of two taken in San Francisco Bay May 1. Note by **B. B. Porter.** <Chicago Field, v. 6, p. 229, May 25, 1878. [F. M.]

California Salmon. [Notice of their ascent up the McCloud and Sacramento rivers in May.] <Forest and Stream, v. 10, p. 350, June 6, 1878.

Salmon canning in Oregon and California. [Editorial. With three woodcuts.] <Forest and Stream, v. 10, p. 398, June 27, 1878.

1878—Another shipment of Shad to California. Notice by **Fred. Mather.** <Chicago Field, v. 9, p. 308, July 6, 1878. [F. M.]

California Salmon in Lake Ontario. [By **John J. Robson.**] <Forest and Stream, v. 10, p. 482, July 25, 1878.

Salmon canning in Alaska. An account of the objections of the Indians to the landing of a lot of Chinese fish canners. From Alaska Cor. "N. Y. Sun." <Chicago Field, v. 9, p. 371, July 27, 1878. [F. M.]

Notes on a Collection of Fishes from Clackamas River, Oregon. By **David S. Jordan,** M. D. <Proc. U. S. Nat. Museum, v. 1, pp. 69–85, Aug., 1878.

The Labrador and Columbia River Fisheries. [From the "New York Sun."] <Forest and Stream, v. 10, p. 507, Aug. 1, 1878.

The Mysterious Salmon. A quotation from the "Astorian" on the subject of the salmon taking the artificial fly, with editorial comment by **Fred. Mather.** < Chicago Field, v. 9, p. 387, Aug. 3, 1878. [F. M.]

The McCloud River Hatchery. [By **K. B. Pratt.**] < Forest and Stream, v. 11, p. 2, Aug. 8, 1878.

Fish Gossip: Abundance of Salmon in the McCloud River, and their annoyance to anglers when fishing for Trout. [Item from "San Francisco Chronicle," with editorial comment by **Fred. Mather.** <Chicago Field, v. 9, p. 403, Aug. 10, 1878. [F. M.]

Gameness of the Quinnat Salmon. [By **Tarleton H. Bean.**] < Chicago Field, v. 10, p. 4, Aug. 17, 1878. [F. M.]

The Fraser River Salmon Season. [From the "New York World."] <Forest and Stream, v. 11, p. 50, Aug. 22, 1878.

Fishing in Northern California. [By **E. J. Hooker.**] <Forest and Stream, v. 11, p. 51, April 22, 1878.

Trout Fishing in Truckee River. Correspondent of the "Sacramento Union." <Chicago Field, v. 10, p. 20, Aug. 24, 1878. [F. M.]

Trouting in Nevada. Catching them in the water-works at Gold Hill and Virginia City. [From "Virginia City Chronicle."] < Chicago Field, v. 10, p. —. Sept. 14, 1878. [F. M.]

Good News from California. [An account of fish-ladders in the Truckee River, from the "Truckee Republican."] <Chicago Field, v. 10, p. 84, Sept. 21, 1878.

Salmon One Cent Each. [Item from Frazer River, from California paper, with editorial comment by **F. Mather.**] <Chicago Field, v. 10, p. 101, Sept. 28, 1878. [F. M.]

Salmon Canning on Columbia River. An account of the process, with statistics. By **Fred. Mather.** <Chicago Field, v. 10, p. 101, Sept. 28, 1878. [F. M.]

Note on the Saurus lucioceps of Ayres. [By **W. N. Lockington.**] <Ann. & Mag. Nat. Hist. (5), v. 2, pp. 348, 349, Oct., 1878.

McCloud River Hatching Station. Daily Record of Salmon taken. [Signed **Livingston Stone.**] < Forest and Stream, v. 11, p. 203, Oct. 10, 1878.

California Trout in New York. [By **Seth Green.**] <Forest and Stream, v. 11, p. 203, Oct. 10, 1878.

McCloud River Hatchery. [Table of Distribution of Salmon Eggs during 1878.] < Forest and Stream, v. 11, p. 222, Oct. 17, 1878.

1878—Land-locking the Quinñat Salmon. Experiment of **H. G. Parker**, Commissioner on Fisheries for Nevada, in Pyramid and Walker Lakes. < Chicago Field, v. 10, p. 165, Oct. 26, 1878. [F. M.]

The Yellowstone as a Trout stream. [*Anon.*] < Forest and Stream, v. 11, p. 263, Oct. 31, 1878.

Another Devil Fish Story. Account of devil-fish (*Ceratoptera*) interfering with a submarine diver, from California paper. < Chicago Field, v. 10, p. 181, Nov. 2, 1878. [F. M.]

Walks around San Francisco. By **W. N. Lockington**. No- III.—Lake Honda and Seal Rock. < Am. Nat., v. 12, pp. 786–793, Dec., 1878. [N. Sp. *Bdellostoma Stoutii*, p. 793.]

Note.—"No. I.—The Ocean Beach" (v. 12, pp. 347–354) and [No. II.—] "The Bay Shore" (v. 12, pp. 505–512) have nothing relative to fishes.

Salmo quinnat in France. [By **Fred. Mather**.] < Forest and Stream, v. 11, p. 360, Dec. 5, 1878. [See, also, pp. 339, 340, Nov. 28, 1878.]

On the occurrence of Stichæus punctatus, (Fabr.) Kröyer, at St. Michæl's, Alaska. By **Tarleton H. Bean**. < Proc. U. S. Nat. Museum, v. 1, pp. 279–281, Dec. 17, 1878.

Report on the collection of Fishes made by Dr. Elliott Coues, U. S. A., in Dakota and Montana during the seasons of 1873 and 1874. By **David S. Jordan**, M. D. < Bull. U. S. Geol. and Geog. Surv. Terr., v. 4. pp. 777–799, Dec. 11, 1878.

Note.—[Contains an "analysis of the genera of American Cyprinidæ, and reference of Pacific slope genera to European types, at pp. 785–790.]

California Salmon in Holland. [Editorial.] < Forest aud Stream, v. 11, p. 420, Dec. 25. 1878.

45th Congress, 3d session. } House of Representatives. { Ex. Doc. 1, pt. 2. Vol. II. | = | Annual Report | of the | Chief of Engineers | to the | Secretary of War | for the | year 1878. | — | In three parts. | — | Part III. | — | Washington : | Government Printing Office. | 1878. |

Appendix NN. | — | Annual Report of Lieutenant **George M. Wheeler**, | Corps of Engineers, for the fiscal year ending | June 30, 1878. [pp. 1421—

Appendix K. | Report upon the Fishes collected during the years 1875, 1876, and 1877, in | California and Nevada, by Prof. **David S. Jordan** and **H. W. Henshaw**. [pp. 1609–1622, pll. 1–4.]

Appendix K 1. | List of Marine Fishes collected on the coast of California near Santa | Barbara in 1875, with notes by Dr. **H. C. Yarrow**, Acting Assistant Surgeon | U. S. A., and **H. W. Henshaw**. [pp. 1623–1627.]

P. 1610, pl. 1, 2, *Catastomus tahoensis* Gill and Jordan.
P. 1610, pl. 3, *Catastomus arœopus* Jordan.
P. 1619, pl. 4, *Salmo Henshawi* Gill and Jordan.

The Sportsman's Gazetteer and General Guide. The Game Animals, Birds, and Fishes of North America : Their Habits and Various Methods of Capture. Copious Instructions in Shooting, Fishing, Taxidermy, Woodcraft, etc. Together with maps. By **Charles Hallock**, Editor of "Forest and Stream"; Author of the "Fishing Tourist," "Camp Life in Florida," etc. Fourth Edition. New York : Forest and Stream Publishing Co. 1878. (12mo.)

1878—Manual of the Vertebrates of the Northern United States, Including the District East of the Mississippi River, and North of North Carolina and Tennessee, exclusive of Marine Species. By **David Starr Jordan**, Ph. D., M. D., Professor of Natural History in Butler University. Second Edition, Revised and Enlarged. Chicago: Jansen, McClurg & Co. 1878. (12mo, 407 pp.)

The Californian Salmon. With an Account of its Introduction into Victoria. By Sir **Samuel Wilson**, Member of the Legislative Council of Victoria. Melbourne: Sands & McDougall, Printers, Collins street West. 1878.

1879.—The Nevada Fish-hatchery. [From Carson City "Appeal."] <Chicago Field, v. 10, p. 332, Jan. 4, 1879. [F. M.]

Capture of a Devil-fish [Ceratoptera]. From California paper. <Chicago Field, v. 10, p. 395, Feb. 1, 1879. [F. M.]

The Fisheries and Other Resources of Alaska. By **H. A. R.** <Chicago Field, v. 10, p. 395, Feb. 1, 1879. [F. M.]

Viviparous Perch [Embiotocidæ. By **Charles Hallock.** From "Sportsman's Gazetteer."] <Forest and Stream, v. 11, p. 513, Jan. 23, 1879.

Fish and Fishing of Oregon. [By **Wm. Lang.**] <Forest and Stream, v. 12, p. 35, Feb. 13, 1879.

Report of the Nevada Fish Commission. [Notice by **Fred. Mather.**] <Chicago Field, v. 11, p. 3, Feb. 15, 1879.

Rapid growth of the Californian Salmon. [*Anon.*] <Forest and Stream, v. 12, p. 55, Feb. 20, 1879.
[An abstract from the "German Fishing Gazette."]

Eastern Trout on the Pacific Slope. [By **H. H. Holt,** Kaloma, W. T. <Forest and Stream, v. 12, p. 105, March 13, 1879.

Rearing Whitefish in confinement. [By **B. B. Redding.**] <Chicago Field, v. 11, pp. 67,68, March 15, 1879.

Interesting Facts from Washington Territory. [By **Chs. Bendire.**] <Forest and Stream, v. 12, p. 154, March 27, 1879.
[Refers to "*Salmo Kennerlyi*", &c.]

The Flounders of our Markets. Read by **W. M. Lockington** before the San Francisco Acad. of Sciences, March 17, 1879. <Scientific Press Supplement, April, 1879; Mining and Scientific Press, April 12 and 19, 1879.

Salmon Fishing in Oregon. [By **H. B.**] <Forest and Stream, v. 12, p. 174, April 3, 1879.

Traits of Rocky Mountain Trout. [By **W. N. Byers.**] <Forest and Stream, v. 12, p. 174, April 3, 1879.

[Notice of a " 'Devil Fish' recently taken on the Pacific coast whose body was four feet long, with a spear-shaped tail and tentacles seven feet long," *i. e.*, a species of Ceratoptera. From the "Santa Barbara Press."] <Chicago Field, v. 11, p. 148, April 19, 1879.

Description of a species of Lycodes (*L. Turneri*) from Alaska, believed to be undescribed. By **Tarleton H. Bean.** <Proc. U. S. Nat. Museum, v. 1, pp. 463-466, April 25, 1879.

The Fishes and Birds of the Pacific Coast. [By **Calamink,** *pseudon* of **John L. Wilson.** <Chicago Field, v. 11, p. 163, April 26, 1879.

[Note relative to the Fisheries of British Columbia. Notice of Report to House of Commons.] <Chicago Field, v. 11, p. 165, April 26, 1879.

1879—Notes on some Fishes of the Coast of California. No. I. By **W. N. Lock-ington.** < Am. Nat., v. 13, pp. 299–308, May, 1879.

California Mountain Trout in Eastern Waters. [By **Seth Green.**] < Forest and Stream, v. 12, p. 264, May 8, 1879.

[See, also, v. 12, p. 288.]

Trout and Salmon Season in California. [*Anon.*] < Forest and Stream, v. 12, p. 277, May 8, 1879.

Angling in California. [Abstract from "Pacific Life."] < Chicago Field, v. 11, pp. 195, 196, May 10, 1879.

[Catfish in California.] < Chicago Field, v. 11, p. 196, May 10, 1879.

Pacific Trout [Salmo iridea] in Eastern Waters. [Note signed **H. W. De Long,** with description appended from Hallock's Sportsman's Gazetteer.] < Forest and Stream, v. 12, p. 288, May 15, 1879.

Does the Western Salmon die after spawning? [By Major, *pseudon.*] < Chicago Field, v. 11, p. 221, May 17, 1879.

California Salmon do not all die after spawning. [By **B. B. Redding.**] < Chicago Field, v. 11, p. 236, May 24, 1879.

The Roe of the Salmon the Indian's Bait. [By **Jonas C.,** Portland, Oregon.] < Chicago Field, v. 11, p. 237, May 24, 1879.

California News. [Notice of expected consignment of eggs from U. S. Commission Fish and Fisheries. *Anon.* From Sacramento "Record-Union."] < Chicago Field, v. 11, p. 244, May 31, 1879.

On a new Genus of Scombridæ. By **W. N. Lockington.** < Proc. Acad. Nat. Sci. Phila. [v. —], pp. 133–136.

[N. g. and sp. *Chriomitra* (p. 153) concolor, p. 134.]

Who branded the Salmon? [Notice of capture of four salmon branded with W. at Westport, Oregon. By **Geo. H. Heather.**] < Chicago Field, v. 11, p. 260, June 7, 1879.

Lake Tahoe. [*Anon.* From "Philadelphia Press."] < Chicago Field, v. 11, p. 260, June 7, 1879.

Grand Success of Shad and Salmon Culture. [By **B. B. Redding.**] < Chicago Field, v. 11, p. 277, June 14, 1879.

Salmon at the Antipodes, being an account of the successful introduction of Salmon and Trout into Australian waters. By Sir **Samuel Wilson,** Member of the Legislative Council of Victoria, [etc.]; author of a work on the Angora Goat, and papers on the Ostrich, the Chinese Yam, etc. London: Edward Stanford, 55, Charing Cross, S. W., 1879. [3d ed., 12°, viii, 252 pp., 1 phot. pl., 1 map folded.]

Partial Contents.

[Introduction dated June 16, 1879.

"The substance of this work, in a slightly different form, under the title of 'The Californian Salmon,' was originally published in the Transactions of the Zoological and Acclimatization Society of Melbourne for the year 1878, and a second small edition was reprinted in Victoria."—From "Preface to the third edition."—See 1878]

1879—The Chinese and other Fishermen of California. [Condensed from San Francisco "Chronicle" by **Fred. Mather.**] < Chicago Field, v. 11, p. 291, June 21, 1879.

On the Occurrence of Hippoglossus vulgaris, Flem., at Unalashka and St. Michael's, Alaska. By **Tarleton H. Bean.** < Proc. U. S. Nat. Museum, v. 2, pp. 63-66, July 1, 1879.

Pacific Coast Shad. [By **William Lang.**] < Forest and Stream, v. 12, p. 457, July 24, 1879.

Notes on New and Rare Fishes. Read before the California Acad. Science by **W. N. Lockington.**] < Scientific Press Supplement, July, 1879; Mining and Scientific Press, Aug. 2 and 16, 1879.

Fish Notes from the Pacific Coast. [By **Robt. E. C. Stearns.**] < Chicago Field, v. 11, p. 389, Aug. 2, 1879.
[Extract from "American Naturalist."]

Curious Facts about Trout [i. e., jumping from flume into water below. By **B. B. R.**, i. e. **B. B. Redding.**] < Chicago Field, v. 11, p. 404, Aug. 9, 1879.

Alaska in Summer.—Second Paper. [By "**Piseco**," i. e. **Lester Beardslee.**] < Forest and Stream, v. 13, p. 553, Aug. 14, 1879.
[Refers, inter alias, to capture and curing of salmon at Port Hunter.]

Largest Salmon on Record. [*Anon.*] < Forest and Stream, v. 13, p. 557, Aug. 14, 1879.
["VICTORIA, June 26.—A salmon that weighed 98 pounds when caught has been received here from the Skeena River Fishery by Mr. Turner, Mayor of Victoria. Its length is 5 feet 11 inches from nose to tail."]

Shad in the Columbia. [By "S."] < Forest and Stream, v. 13, p. 585, Aug. 28, 1879.
[Refers probably to *Pomolobus*.]

Trolling for Salmon. [*Anon.*] < Forest and Stream, v. 13, p. 588, Aug. 28, 1879.
[Relates to Columbia River.]

Oregon. [Record of a trout-fishing expedition. By **William Lang.**] < Forest and Stream, v. 13, p. 589, Aug. 28, 1879.

The McCloud River Fishery. [*Anon.*] < Forest and Stream, v. 13, p. 604, Sept. 4, 1879.

Salmon a Nuisance to Trout Fishers. [*Anon.* By **Fred. Mather.**] < Chicago Field, v. 12, p. 52, Sept. 6, 1879.

The North Pacific Codfishery. [By **W. N. Lockington.** Reprinted from "Pacific Life."] < Chicago Field, v. 12, p. 53, Sept. 6, 1879.

[Notice of Trout passing through flume under pressure of 376 pounds to the square inch. *Anon.*] < Chicago Field, v. 12, p. 53, Sept. 6, 1879.

[Notice of Catfish—Amiurus albidus?—5 to 15 inches long, taken in Sausal Lagoon, where planted three years before. *Anon.*] < Chicago Field, v. 12, p. 53, Sept. 6, 1879.

The Pacific Salmon Fisheries. [*Anon.*] < Chicago Field, v. 12, p. 69, Sept. 13, 1879.

[Notice of Catfish—Amiurus albidus?—taken in McCloud's Lake, Stockton. *Anon.*] < Chicago Field, v. 12, p. 69, Sept. 13, 1879.

The Trans-Continental Expedition of the California Fish Commissioners. [By H. A. L.] < Forest and Stream, v. 13, p. 645 (3 col.), Sept. 18, 1879.

1879—Review of the Pleuronectidæ of San Francisco. By **W. N. Lockington.**
< Proc. U. S. Nat. Museum, v. 2, pp. 69–96, July 2—Sept. 19, 1879.
[N. sp. *Hippoglossoides Jordani*, p. 73; *Glyptocephalus Pacificus*, p. 86; *Glyptocephalus zachirus*, p. 88.

[Notice of Catfish for Susan River and Eel Lake. *Anon.*] < Chicago Field,
v. 12, p. 85, Sept. 20, 1879.

The first biennial report of the Nevada Commission. [Notice by **Fred. Mather.**] < Chicago Field, v. 12, p. 85, Sept. 20, 1879.

Habits of California River Salmon. [*Anon.* Extract from "Sacramento
Bee."] < Chicago Field, v. 12, p. 100, Sept. 27, 1879.

Fish Culture Operations in California. [By **Livingston Stone.**] < Forest
and Stream, v. 13, p. 685, Oct. 2, 1879.
[Refers to Salmon.]

Why Salmo Quinnat does not take the Fly. [*Anon.* by **Charles Hallock.**
< Forest and Stream, v. 13, p. 685, Oct. 2, 1879.

Washington Territory. [By "MULTNOMAH," *pseudon.*] < Forest and Stream,
v. 13, p. 687, Oct. 2, 1879.
[Relates to fishing in "the great Spokane country."]

Salmon Fishing on the Pacific. [Incomplete. By C. R.] < Forest and
Stream, v. 13, p. 689, Oct. 2, 1879.

The Fishery of Mr. A. P. Rockwood [near Salt Lake City. *Anon.* From "The
Juvenile Instructor."] < Chicago Field, v. 12, p. 115, Oct. 4, 1879.

Do Fish hear? [By **W. N. Lockington.** From "Pacific Life."] < Chicago
Field, v. 12, p. 116, Oct. 4, 1879.

Trout in the Truckee. [*Anon.* From "Sacramento Bee."] < Chicago Field,
v. 12, p. 117, Oct. 4, 1879.

California. [Record of good Grilse-fishing in September.] By **B. B. Redding**
< Forest and Stream, v. 13, p. 715, Oct. 9, 1878.

The Game and Fish of Alaska. [By "PISECO," *i. e.* **Lester Beardslee**, U. S.
N.] < Forest and Stream, v. 13, pp. 723–724, Oct. 16, 1879.

Salmon Eggs from the Pacific. [By **Livingston Stone.**] < Forest and
Stream, v. 13, p. 725, Oct. 16, 1879.

California Fishing. [By **E. J. Hooper.**] < Forest and Stream, v. 13, p. 728
Oct. 16, 1879.

Wyoming Territory. [Note on Trout-fishing. By "MULTNOMAH," *pseudon.*]
< Forest and Stream, v. 13, p. 728, Oct. 16, 1879.

Spawn in off season [of Californian Trout. By **E. C. Tallant.** With editorial note.] < Forest and Stream, v. 13, p. 744, Oct. 23, 1879.

The Redfish of the Northwest. [By **Ch. Bendire.** With editorial note.]
< Forest and Stream, v. 13, p. 745, Oct. 23, 1879.

Rocky Mountain Trout. [By FLYFISHER, *pseudon.*, **J. J. Stranahan**, Chagrin
Falls, O.] < Chicago Field, v. 12, p. 164, Oct. 25, 1879.

"Mountain Trout".—(Salmo virginalis). [By **Gordon Lamb.**] < Chicago
Field, v. 12, p. 164, Oct. 25, 1879.

Fishing in Gray's Harbor [*i. e.* Salmon-fishery. *Anon.* From "Olympia
(Washington Terr.) Transcript." < Chicago Field, v. 12, pp. 164, 165, Oct.
25. 1879.

1879—Codfishing in the Pacific. [*Anon.*] From "San Francisco Alta.") <Chicago Field, v. 12, p. 165, Oct. 25, 1879.

California Trout in New York State. [By **Clarence A. Farnum.**] < Forest and Stream, v. 13, p. 765, Oct. 30, 1879.

Salmon Fishing on the Pacific. [By C. R.] < Forest and Stream, v. 13, p. 767, Oct. 30, 1879.

Why Salmo Quinnat does not take the Fly. [Editorial.] < Forest and Stream, v. 13, p. 770, Oct. 30, 1879.

Notes on Pacific Coast Fishes and Fisheries. By **W. N. Lockington.** < Am. Nat., v. 13, pp. 684–687, Nov., 1879.

Notes on some undescribed Fishes of the Pacific Coast. By **W. N. Lockington.** < Scientific Press Supplement, v. —, p. 76, Nov., 1879.

Carp Breeding in California. [*Anon.* From "Sonoma Index."] < Chicago Field, v. 12, p. 180, Nov. 1, 1879.

Trout Culture in Nevada. [*Anon.* From "Virginia City Enterprise."] < Chicago Field, v. 12, p. 180, Nov. 1, 1879.

Fish in Washington Territory. [*Anon.* From the "Experiment."] < Chicago Field, v. 12, p. 180, Nov. 1, 1879.

Washington Territory. [Abundance of Trout. By MULTNOMAH, *pseudon.*] < Forest and Stream, v. 13, pp. 795–796, Nov. 6, 1879.

The Redfish of Idaho. By **Charles Bendire.** < Forest and Stream, v. 13, p. 806, with fig., Nov. 13, 1879.

[The figure appears to represent *Hysifario kennerlyi.*]

California Notes. (From the "San Francisco Bee.") < Chicago Field, v. 12, p. 213, Nov. 15, 1879.

Some Fishes of Oregon. By **C. J. Smith.** < Forest and Stream, v. 13, p. 826, Nov. 20, 1879.

The Trout of Utah. [Notice of its rising to a fly.] By **C. B. Western** < Forest and Stream, v. 13, p. 826, Nov. 20, 1879.

California Fishing Notes. [From "Sacramento Bee."] < Chicago Field, v. 12, p. 229, Nov. 22, 1879.

California Fishes. By **B. B. Redding.** < Forest and Stream, v. 13, p. 847 Nov. 27, 1879.

Mountain Trout.—*Salmo virginalis.* By **Gordon Land.** < Chicago Field, v. 12, p. 245, Nov. 29, 1879.

The Fishes of Klamath Lake, Oregon. By **E. D. Cope.** < Am. Nat., v. 13, pp. 784–785, Dec., 1879.

[N. sp. *Chasmistes luxatus* (p. 784); *Chasmistes brevirostris* (p. 785); ? *Mylopharodon* sp. (785).]

Annual Record of Science and Industry for 1878. | Edited by **Spencer F. Baird** with the assistance of eminent men of science. | New York: | Harper & Brothers, Publishers, Franklin Square. 1879. [12°.]

The North American Trout and Salmon. pp. 467–470.

Ichthyologische Beiträge (VIII). Von Dr. **Franz Steindachner.** < Sitzb. K. Akad. Wissensch., B. 80, Abth. i, pp. . ("Juli-heft.") [Author's extra, received by mail Oct. 22, 1879.]

[N. sp. *Corvina (Johnius) Jacobi*, San Diego, p. 3; n. g. and sp. *Typhlogobius californiensis*, San Diego, p. 24; and *Gobius Nowberrii*, p. 17, *Engraulis ringens*, p. 62, also commented upon]

INDEX.